WBI Development Studies

Building Knowledge Economies

Advanced Strategies for Development

The World Bank
Washington, D.C.

1818 H Street, NW
Washington, DC 20433
Telephone: 202-473-1000
Internet: www.worldbank.org
E-mail: feedback@worldbank.org

2 3 4 10 09 08

This volume is a product of the staff of the International Bank for Reconstruction and Development / The World Bank. The findings, interpretations, and conclusions expressed in this volume do not necessarily reflect the views of the Executive Directors of The World Bank or the governments they represent.

The World Bank does not guarantee the accuracy of the data included in this work. The boundaries, colors, denominations, and other information shown on any map in this work do not imply any judgement on the part of The World Bank concerning the legal status of any territory or the endorsement or acceptance of such boundaries.

DOI: 10.1596/ 978-0-8213-6957-9

Library of Congress Cataloging-in-Publication Data

Building knowledge economies : advanced strategies for development.
p. cm.
"World Bank Institute, Knowledge for Development Program."
ISBN-13: 978-0-8213-6957-9
ISBN-10: 0-8213-6957-1
ISBN-13: 978-0-8213-6958-6 (electronic)
1. Information technology--Economic aspects. 2. Technological innovations--Economic aspects. 3. Knowledge management. I. World Bank Institute. Knowledge for Development Program.
HC79.I55B847 2007
303.48'33--dc22

2007018122

Contents

Boxes

Figures

Tables

Foreword

Knowledge has always been an essential force in economic development. But in today's increasingly knowledge-based world, more and more countries are embracing knowledge and innovation-related policies to spur growth and competitiveness. At the same time, because their institutions are weak, many developing countries are struggling to find ways to produce relevant knowledge and transform it into wealth, as well as to adapt and disseminate existing knowledge for their development.

Produced by the World Bank Institute's Knowledge for Development (K4D) program, *Building Knowledge Economies: Advanced Strategies for Development* brings together several years of analytical work and research on the knowledge economy in client countries. Its premise, supported by case studies and a great deal of data, is that countries at all levels of development should consider embarking on a knowledge- and innovation-based development process. Now more than ever, grey matter is every country's main sustainable and renewable resource. The book therefore deals with advanced development strategies that focus on policies to identify, enhance, and exploit intangible assets in the areas of education, innovation, information and communication technology (ICT), and the prerequisite economic and institutional regime. Intended primarily for the policy-making community in developing countries, the book provides policy directions that can help political, social, and economic leaders embark on a knowledge-based development process.

In an ambitious but self-effacing manner, the authors challenge policy makers to adopt new policy agendas and mindsets to help improve national growth, competitiveness, and economic welfare in a rapidly connected and globalized world. Because up to 20 years may be needed to reap the fruits of educational reforms (even as investments in information and communications technology and improvements to the business climate show their beneficial effects much sooner), the authors take a long-term perspective. And because every society is different, they do not provide recipes. But they do advocate adapting what has worked elsewhere. The best practices held up for adaptation have emerged largely from the experience of semi-industrialized and developing economies that have conceived and implemented pragmatic policy agendas to put knowledge to work, often with the benefit of analytical and conceptual work at the World Bank and elsewhere on aspects of the knowledge economy. That work, too, is reviewed here.

Building Knowledge Economies: Advanced Strategies for Development is not about the ICT revolution or high technology industries. Rather, it is about economic development as a process of generating relevant knowledge and putting that knowledge to work to generate further growth. As such, it is about the transformation of countries in an open-ended process that is defined more by the initial, necessarily messy, conditions than by a desired end-state—a specification that applies to most countries at all income levels. I hope that this book can be useful in providing some insights into how to build development strategies adapted to the challenges and opportunities of our times.

Frannie A. Léautier
Vice President
World Bank Institute

Acknowledgments

This report was prepared by the World Bank Institute's Knowledge for Development (K4D) team, led by Jean-Eric Aubert. The team included Derek Hung Chiat Chen, Ronald Kim, Yevgeny Kuznetzov, Kurt Larsen, Anuja Utz, Douglas Zhihua Zeng, and Justine White (who coordinated the many inputs to the report). Contributions were also provided by Julio Aguirre, Do-Geol Ahn, Capucine Edou, Aisling Quirke, Alexey Volynets, and Pierre Sauvé (consultant).

We are grateful to Professor Carl Dahlman of Georgetown University and Jean-Francois Rischard, a retired vice president of the World Bank, who provided detailed comments and invaluable suggestions on earlier drafts. We also benefited from the comments of World Bank peer reviewers Severin Kodderitzsch (Environmental and Social Sustainable Development Network), Brian Levy (Europe and Central Asia Region), Jamil Salmi (Human Development Network), and Shahid Yusuf (Development Research Group), and of several external reviewers, including Luc Soete (Director UNU-MERIT, the Netherlands).

Frannie Léautier, former vice president of the World Bank Institute, deserves our special thanks for her generous and unstinting advice and support in the preparation of the report. We are grateful for the same reason to Bruno André Laporte, WBI's manager for knowledge and human development .

Doranne Lecercle and Steven Kennedy edited the report, and Tomoko Hirata designed the cover. Much help and support was received by program assistants Faythe Calandra and Michelle Lemaire, to whom the authors extend their thanks. Anuja Utz and John Didier oversaw production and shepherded the book through the publication process.

Jean-Eric Aubert

Introduction

Knowledge—The Heart and Mind of Economic Development

In the Mauritanian desert, a group of tourists is guided by young English-speaking, formerly unemployed graduates. They draw on the indigenous knowledge of nomads to help explain the movements of dunes, meteorological change, and archeological remains. Owing to a very efficient telecommunication policy implemented by the government, tourists are accommodated in hotels with rapid Internet connections, and this has been an important factor in their decision to come to Mauritania. Along with tourism, new activities flourish: garages for repairing the four-wheel drive vehicles that transport visitors, cultural events organized in places with ancestral traditions. These initiatives are supported by foreign investors attracted by an improved business climate.

Indian entrepreneurs in an office in Bangalore stay in contact with colleagues based in Silicon Valley; they are working jointly on a new software product. Meanwhile, down the road, teleworkers are handling customer-service calls from the United Kingdom. Their fluency in English is just one of the reasons why the U.K. firm decided to invest there. High levels of computer literacy and other competencies among workers, combined with low labor costs and a reliable phone network, made it an extremely attractive investment. Further down the road, an Indian pharmaceutical company is developing new medicines based on traditional remedies and the latest scientific advances. Following in Bangalore's path, the cities of Chennai, Mumbai, and Hyderabad, among others, have created hundreds of thousands of new jobs for skilled knowledge workers. The new opportunities have reduced poverty and raised living standards across entire regions; the forces behind them have developed a spectacular momentum that is challenging the global economy.

Thus, knowledge is being put to work to accelerate and deepen the development process. Applied to all types of innovation, including the more modest ones in the use of basic technology, it becomes a major resource for generating wealth and jobs. Clearly, various types of knowledge, including the most traditional, can be of use; at the same time there is a need to invest in the most advanced technologies to be part of the global economy.

This book argues that whatever their level of development, countries should consider embarking on a knowledge- and innovation-based development process. In these times of accelerated globalization, "grey matter" is a country's main durable resource. Its exploitation for economic and social well-being is increasingly at the center of development strategies.

The central role of knowledge and innovation in economic growth is widely acknowledged in advanced countries, and the experience of those that have championed this new paradigm has led to the coining of the expression "knowledge economies" (KEs). However, this notion is less widely accepted among developing countries, and the elaboration of appropriate and efficient policies for knowledge-based development processes is particularly challenging in view of the more difficult political and economic environment in many of these countries. This book aims at casting some light on these issues.

This book—whose *intended audience is primarily the policy-making community of developing countries*—deals with advanced development strategies that focus on policies to address intangibles: education, innovation, information and communication technology (ICT), and the prerequisite economic and institutional regime.

Main Messages

Message 1. Knowledge and innovation have played a crucial role in development from the beginnings of human history. But with globalization and the technological revolution of the last few decades, knowledge has clearly become the key driver of competitiveness and is now profoundly reshaping the patterns of the world's economic growth and activity. Both developed and developing countries should therefore think, with some urgency, about their future under a KE heading.

Message 2. To become successful knowledge economies, countries have to rethink and act simultaneously on their education base, their innovation systems, and their ICT infrastructure, while also building a high-quality economic and institutional regime. Policies for these four pillars have to reflect the country's level of development and will often have to be gradual. However, experience shows that some successful KE champions have been able to achieve spectacular leaps forward within a decade.

Message 3. Many if not most of the countries that have made rapid progress have staged nationwide KE-inspired programs of change. Such programs have been pragmatic and country-specific, yet some common points emerge: the need to promote trust and societal cohesion around the KE program; the need to work at the four pillars through a combination of top-down reforms and bottom-up initiatives; and the need for a well-communicated KE vision.

Structure

In the first three chapters we respond to three basic questions of interest to policy makers in developing countries: Why should we invest in a knowledge economy? What policy steps are needed to build one? And how do we get from here to there?

Chapter 1 (the "why" chapter), describes the knowledge revolution, which is leading us into a postindustrial age in which brains, not brawn, are the best means of coping with intensified competition and new challenges, including those related to human development and the global environment. In explaining the foundations and the model of knowledge economies, chapter 2 (the "what" chapter) outlines the four KE pillars, provides elements of our benchmarking methodology, and relates KE achievements to recorded growth and competitiveness. To understand the KE process, chapter 3 (the "how" chapter) starts with the examples of three KE champions, Finland, Ireland, and the Republic of Korea. While these are now advanced economies, they offer a few generally applicable lessons of determination and pragmatism. We also address how to implement change through steps both incremental and radical, and through a mixture of bottom-up and top-down initiatives, highlighting the importance of adapting policy actions to countries' circumstances—not only their development levels and growth patterns, but also their sociocultural characteristics.

The ensuing chapters examine and document in detail the four KE policy pillars. Chapter 4 on the economic and institutional regime discusses governance, regulation, finance, and trade as they bear on the KE development process. Chapter 5 on innovation addresses the question of support for innovators, the strengthening of research and technology infrastructure, the diffusion of basic technologies, and the promotion of specific industries. Chapter 6 deals with the national ICT infrastructure, addressing related applications, institutions, and regulations, as well as access to that infrastructure and the development of the skills needed to build, maintain, and use it. Chapter 7 looks at primary and secondary education, higher education, and lifelong learning from a KE perspective, providing insights on the mobility of human resources and the role of diasporas. Chapter 8 discusses policy agendas for selected sets of countries. We begin with the two emerging giants—China and India. Two exemplary groups of middle-income countries are then discussed. Finally the focus turns to low-income countries, particularly in Sub-Saharan Africa.

Throughout the book, we sketch out policy principles adapted to different levels of development (that is, in high-, middle-, and low-income countries) and point out opportunities for countries to make rapid progress by "leapfrogging." Examples from various parts of the world are provided as possible sources of inspiration for policy-making communities. We sum up the main lessons of our investigation in a conclusion entitled "Moving Ahead to a Knowledge Economy."

Information and Analytical Bases

The analysis and information on which this book is based are largely drawn from work by the World Bank Institute's Knowledge for Development program, launched in 1999, which has carried out a number of KE case studies and organized KE forums in different parts of the world. Significant use is also made of the methodology and database developed by the program for appraising countries through the KE lens. The "Knowledge Assessment Methodology" is presented in an annex to chapter 2. More information is available at http://www.worldbank.org/kam.

We also consulted studies carried out by units of the World Bank on knowledge-related issues and policies (education, ICT, and so on) and on specific regions or countries, with a focus on knowledge and innovation strategies. A review of previous World Bank work on the knowledge economy forms the appendix to this volume.

* * *

This book is both an ambitious and a humble exercise. It is ambitious in that it promotes a new type of development strategy, covering a broad and diversified range of policies focused on the intangible ingredients of the economic growth process. It is humble because it does not pretend to be more than an introduction to a very complex subject. In pointing out promising paths, we make no claim to exhaustiveness in our treatment of country cases, or to detailed prescriptions in the policy orientations suggested.

1

Toward Knowledge Economies: Global Trends and Issues

This chapter recalls the crucial role of knowledge in national development throughout history, particularly as a foundation of economic and social development. It then turns to the knowledge revolution now driving the transition to a postindustrial, globalized economy. Finally, it discusses the challenges and opportunities faced by both industrialized and developing countries.

A Brief Historical Retrospective

From Prehistoric to Industrial Times

Knowledge has been of decisive importance in mankind's development. Early man's ability to make fire was a tremendous advance transmitted within and among tribes. Later, primitive societies accumulated knowledge about plants, animals, and minerals essential to their survival for thousands of years. Aspects of this knowledge are still of fundamental importance today in the fields of health care and nutrition, with applications in modern medicine.[1]

The Neolithic period (10,000–5,000 B.C.) brought more productive agricultural practices to various parts of the world—Mesopotamia, southern China, Central America, and West Africa—owing to advances in plant seeding, plowing techniques, metallurgy, tool making, and so on. This knowledge was gradually expanded and stored in written form to be transmitted from generation to generation, forming the basis of major civilizations into the Middle Ages. Indeed, it continued to support the development of society up to the Industrial Revolution.

The Industrial Revolution took shape in Europe following the rediscovery of ancient Greek knowledge during the Renaissance, and the resulting Scientific Revolution. The Greeks had developed an investigative approach to nature that led to important discoveries in astronomy, physics, and mathematics—with the Romans later adding important engineering techniques. When Europe fell into centuries of darkness after the fall of the Roman Empire, this collective knowledge was kept alive by the Arabs—along with accumulated knowledge from China and India—and by Irish monks who preserved and copied the works of antiquity. It was on this foundation that Europe reemerged in the late Middle Ages to give birth to the Renaissance explosion in science, art, and other areas of knowledge development.

1. The Indigenous Knowledge Program of the World Bank's Africa Region is a good source of information on this subject. http://www.worldbank.org/afr/ik/.

China had produced some important technologies, at times in advance of Europe, but the country stagnated and failed to make the leap into the Industrial Revolution.

The Industrial Revolution began with the development of the steam engine and textiles machinery. It linked entrepreneurial dynamism with sources of knowledge and invention. Britain took advantage of this relation to gain a position of leadership, with other countries in tow (Landes 1999). Germany strengthened the links between educational and industrial organizations during the second industrial revolution (involving chemicals and mechanics), to become the technological leader among industrializing nations. The United States benefited from these achievements, with large investments in public education and modern factory and assembly processes—enabled by the electricity supply infrastructure crucial to the nation's successful and rapid development in the late 19th through the 20th century.

Industrialization continued to spread around the world, taking root in the countries best equipped to absorb it. The former Soviet Union benefited from massive efforts in education and science—and developed a very large technological system—but disappeared 70 years later, the victim of the intrinsic inefficiency of its economy.

In Asia, Japan adopted Western technologies to build a modern economy. Its efforts to further national education bore fruit after the Second World War, when it based its reconstruction on an active export strategy. As the level of Japan's technology gradually increased, it upgraded its research and development (R&D) system, becoming a world leader in the mass production of cars, electronics, and other high-tech equipment. The Republic of Korea and Taiwan (China) followed a similar path.

Recent Developments

Since the invention of the transistor, and throughout the stages of the microelectronic revolution, there has been rapid progress in information processing. With information-processing capability doubling every 24 months according to Moore's law,[2] production is increasingly automated. An increase in computer speed and processing power has facilitated other technological leaps. Developments in genetics and associated biotechnologies have spurred tremendous progress in the life sciences and their applications. Advances in the manipulation of matter have led to the invention of new materials and nanotechnologies. Finally, important new developments have occurred in the field of energy. As we probe more deeply into the understanding of time, life, matter, and energy, all these technological changes tend to converge in a way that transforms the entire economic and social system (figure 1.1; Gaudin and Dégremont 1993).

Technical advances permeate our daily lives. The examples are numerous—from electronics in cars, new shoe glues, and reservations made over the Internet, to express shipments and the ubiquity of computers in offices and homes. Such technological progress has induced a series of crucial changes. For example, advances in microelectronics and telecommunications have had a profound global impact in an extraordinarily short period of time (figure 1.2), while the full impact of nanotechnology has yet to be seen.

2. Moore's law is the empirical observation that the transistor density of integrated circuits doubles every 24 months with respect to minimum component cost. It is attributed to Gordon Moore (1965), a cofounder of Intel.

Figure 1.1 Time, Substance, Life, and Energy: An Evolution Through Time

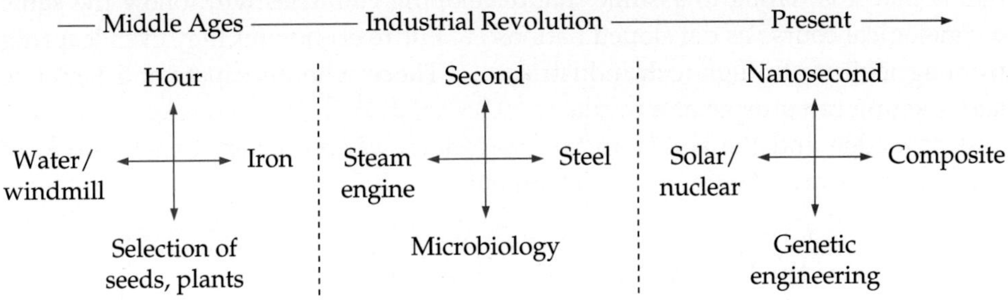

Source: Gaudin and Dégremont 1993.

Figure 1.2 World Growth in Key Information and Communication Technologies, 1991–2004

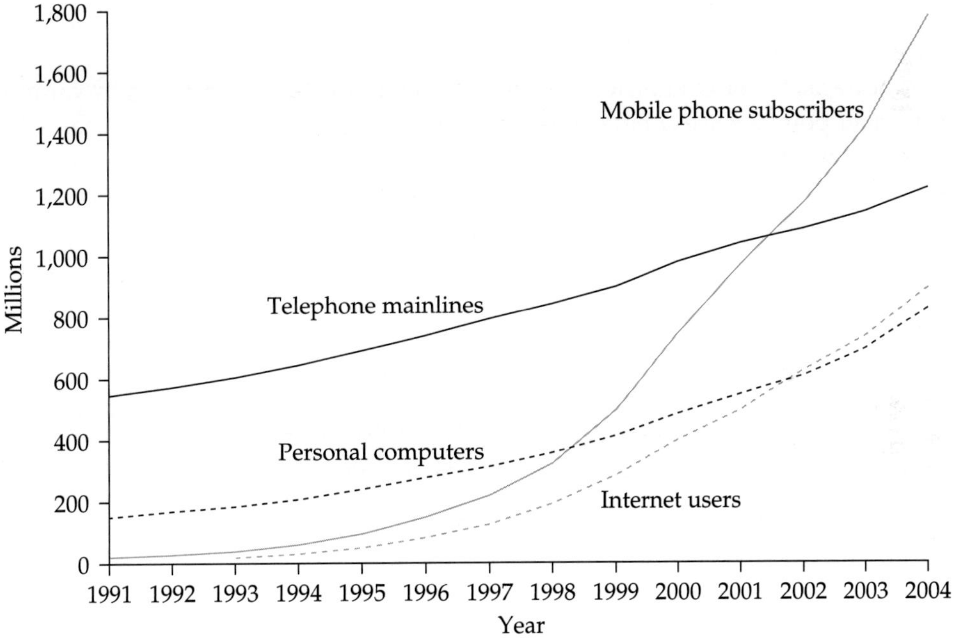

Source: Authors' calculations from World Bank internal data.

Rapid advances in telecommunications and transport have accelerated globalization, creating a new playing field for worldwide competition with considerable opportunities for agile countries. Changes to the world economy will be discussed later, though it is worth noting the stellar performances of small advanced economies such as Singapore and Ireland; the growth of U.S. productivity (much of it owing to the ability of established sectors to benefit from globalized economies' dependence on logistics) and the stalling of Europe; the explosion of offshoring and outsourcing, including China's emergence as the world's assembly line and pockets of amazing IT-based growth in cities such as Bangalore; and dramatic leaps in teledensity—even in Africa. The least developed countries (LDCs) are growing at 6 percent on average—with an unprecedented world growth rate of 5 percent.

As stated in a recent article in the *Economist*,[3] "the lesson to be drawn from all this is that it is wrong to assume that developing countries will follow the same technological course as developed nations . . . Entire economies may even leapfrog from agriculture to high-tech industries . . . Those who anticipate and facilitate leapfrogging can prosper as a result."

Knowledge and the ability to take advantage of technology are—more than ever—at the heart of successful development.

Knowledge as the Foundation of Development

To illustrate the dramatic role played by knowledge in the development process, an important econometric study conducted by the World Bank (1999) compared the per capita gross domestic product (GDP) growth profiles of Ghana and the Republic of Korea over a half-century (figure 1.3). It estimated that close to two-thirds of the differences between the two were attributable, not to the accumulation of physical capital and labor, but to other sources of growth and productivity in which knowledge was crucial, though difficult to estimate (see the discussion of total factor productivity below).

The decisive role of knowledge in development can also be illustrated by various threshold points. For example, it is generally estimated that sustainable economic takeoff cannot take place below a threshold literacy rate of 40 percent and a minimum telephone density of 30 percent (World Bank 2003c). A number of developing countries have yet to reach that threshold.

Figure 1.3 *GDP Growth in Republic of Korea and Ghana over 50 years*

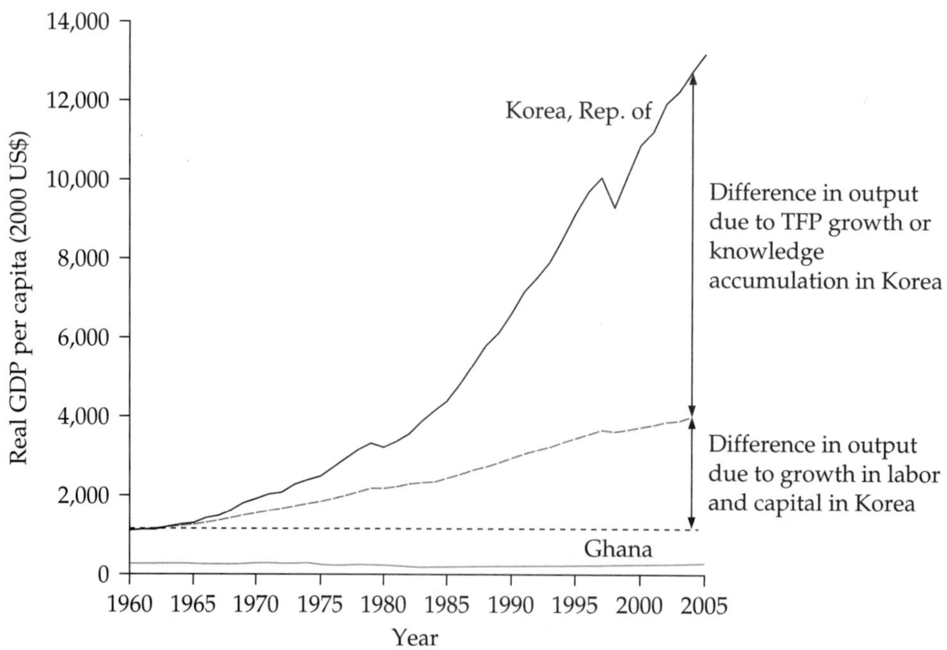

Source: Authors' calculations based on World Bank internal data.

3. "Behind the Bleeding Edge: Technology Leapfrogs," *The Economist*, September 23, 2006.

The capacity to use knowledge effectively allows individuals, enterprises, and communities to utilize resources and improve their well-being, thereby contributing to development. This is by no means a straightforward process. Simple exposure to knowledge, while necessary, does not ensure its effective use. One must be able to select the right form of knowledge, master its application, adapt it to specific circumstances, keep up with changes, and make improvements.

Knowledge can take many forms: tacit or codified, technical or societal. Among its many contributions to economic and social development, three are briefly sketched out below: knowledge as the (a) driver of competitiveness and productivity, (b) facilitator of welfare and environment, and (c) enabler of institutions and governance.

Knowledge as the Driver of Competitiveness and Productivity

When a firm, industry, or country acquires a competitive edge, knowledge becomes the basis of a "rent" (income over and above normal profit) that replaces the rents derived earlier from resources and cheap labor. Examples are many, from textiles and clothing to software (table 1.1).

Knowledge is special because it is difficult to obtain, whether through creation or purchase. Unlike information, knowledge involves combinations of facts that interact in intangible ways. Because it is difficult to obtain, it constitutes an entry barrier to growth—and this entry barrier, in turn, helps generate the rent earned from knowledge. There are several types of knowledge rent: technological (control of scarce process or product capabilities), human resources (availability of unique or advanced human skills and know-how), organizational (control of unique or advanced management practices), and marketing and design (both increasingly important in recent years, with a direct correlation to consumer know-how).[4] Knowledge rents such as these are transitory, however, and require continuous renewal.

Table 1.1 *Knowledge and Natural Resource Rents*

Type of rent	*Previous areas of rent*	*New and emerging areas of rent*
Natural resources	High-grade copper deposits	High-grade platinum deposits
Knowledge		
Technology	Copy lathes Internal combustion engines	Computer-aided design Fuel cells Biotechnology applications
Human resources	Tool-making artisans	Software engineers
Organizational	Mass production organized by managers, quality inspectors	Continuous learning management of just-in-time supply chains, single-unit flow, and "quality at source"
Marketing and design	Levi-Strauss	The Gap
Relational	Short-term and arms-length buyer-supplier interactions	Long-term, discussion-rich relationships within supply chains

Source: Adapted from Kaplinsky 2005.

4. The discussion of knowledge rents follows Kaplinsky 2005.

As is clear from the examples in table 1.1, all sectors—from the most traditional to those on the cutting edge—can benefit from knowledge advances.

Knowledge influences competitiveness, economic growth, and development as long as it finds concrete applications—in other words, as long as it is at work. Since the emergence of classic economic theory in the 18th century, economists have sought to determine the sources of economic growth, from Adam Smith's examination of the division of labor in *The Wealth of Nations,* to Joseph Schumpeter's analysis of the importance of innovation in capitalism in the mid–20th century. After the Second World War, Robert Solow's (1956) approach offered a unified analytical framework, according to which economic growth is ultimately determined by extra-economic, exogenous factors such as technological progress.

The need to assess the importance of technological progress for sustainable growth, including related investments such as education, created the impetus for new growth theories that have tried to endogenize (or "include") technological progress in their models (box 1.1). These new theories and related econometric models help explain why countries' economic trajectories tend to diverge (box 1.2), and

***Box 1.1** Theories of Growth: Integrating Knowledge in Economic Models*

A neoclassical model of long-term economic growth was first developed by Robert Solow in a seminal 1956 paper, "A Contribution to the Theory of Economic Growth." Within his analytical framework, production levels in any given period depend on the quantity of labor and capital in use, but the economy's long-term growth rate depends on birth rates and labor productivity. This model proposes a rather optimistic view of development by stating that countries sharing similar features can converge if two fundamental assumptions hold: (a) the decreasing marginal productivity of production factors and (b) constant economies of scale. In this context, economic stagnation appears inescapable in the long term, and assumed exogenous technological progress is the only mechanism that makes a constant and positive growth rate possible. There are at least two important limits to Solow's growth theory, however. First, it announces a convergence of the world's economies that has not occurred. Second, on a theoretical level, the assumption that technological progress is an exogenous factor cannot be accounted for in economic terms.

Faced with these shortcomings, economists have sought to endogenize technological progress—also called total factor productivity—in their economic growth models. In this view, progress is no longer considered a natural phenomenon spurred by agents' behavior and extra-economic incentives, but as an economic one subject to specific choices. This also calls into question the decreasing marginal productivity of production factors. Indeed, such factors can be considered constant in endogenous growth theories because the efficiency of the production process can be continuously improved. This is mainly because positive externalities may result from the interaction of firms, or from the public goods at their disposal. In Paul Romer's canonical model of 1986, externalities appear because firms' investments not only benefit specific agents, but also enhance global productivity by raising the level of technological development in the economy as a whole. Introducing the notion of human capital, in 1988 Robert Lucas pointed out the positive externalities that may result from the simple interaction of skilled agents and their competent counterparts. From another viewpoint, Robert Barro's 1990 model emphasizes the effect of public spending on infrastructure, and of a stable regulatory environment on raising the overall growth rate of production and income per worker.

Source: Guellec and Ralle 2003.

Box 1.2 Why Growth Performances Differ: Total Factor Productivity

Differences in income and growth rates are explained less by differences in capital accumulation—both physical and human—than by total factor productivity. Empirical research conducted by the World Bank, based on production function estimates for a large sample of countries, shows that much of the growth rate is to be attributed to TFP. The effect increases if improved human efficiency is added.

Variance decomposition	*Contribution of*	
	TFP growth	Capital growth
Without human capital (60 countries, non-oil exporters)		
1960–92	0.58	0.41
1980–92	0.65	0.21
With human capital		
1960–92 (44 countries)	0.94	0.52
1980–88 (50 countries)	0.68	0.20

Source: Easterly and Levine 2001.

therefore help justify government action and investment in public goods such as education and infrastructure, which facilitate the use of knowledge and innovation.

Using these growth models, however, it continues to be hard to measure the effect of knowledge on economic growth. For example, it is difficult to single out the contribution of knowledge to total factor productivity, which is at the heart of the growth process and can be affected by other parameters such as the more effective utilization of human or physical capital.

In contrast, a more disaggregated set of analytical frameworks and indicators—the knowledge assessment methodology set forth in this book to appraise knowledge-based economies—can cast new light on knowledge-based economic growth (see chapter 2).

Knowledge as the Facilitator of Welfare and Environmental Stewardship

Knowledge improves nutrition, cures epidemics, and protects against natural dangers.

KNOWLEDGE NURTURES. The "green revolution" is a striking example. This is the term coined for the increased food production resulting from improved strains of wheat, maize, rice, and other cereals developed by Norman Borlaug and others in the 1960s and diffused throughout the world. It was particularly successful in India, where scientific advances and dissemination efforts throughout villages were effectively linked (box 1.3). The initial focus was on the creation, dissemination, and adaptation of better grains and relevant agricultural know-how. However, the revolution's potential could not be realized until millions of farmers were convinced that they could benefit from using the new grains, and, more importantly, could obtain the credits to buy them. The two ends of the knowledge spectrum—exploration of the new, and diffusion and use of the known—were integrated. This is consistently important for knowledge-based development in both industrialized and developing countries.

Box 1.3 *The Green Revolution in India*

Before India's independence in 1947, its history was studded with periods of famine, drought, and food shortages. From 1850, at least 20 million lives were lost in some 20 occurrences of famine. The huge gap between minimum requirements and food supply continued throughout the post war period, making India reliant on food imports to meet domestic requirements. Realizing the disastrous consequences of the increasing gap between population growth and food production, in the 1960s the government took stock and launched a number of initiatives. These included: (a) a vigorous "grow more food" campaign aimed at attaining agricultural self-sufficiency; (b) agricultural universities set up in all regions to carry out research in the sector; and (c) the adoption of the U.S. land-grant model of agricultural research, which links research with implementation. The Indian government sought to ensure that university research would go from the laboratories to the farms. Irrigation modernization projects were also undertaken, lowering peasants' vulnerability and reliance on monsoon rain. With this basic framework in place, the multiplier effect was provided by new hybrid strains of wheat developed by Norman Borlaug, which were tested by researchers at Indian universities and found to significantly increase productivity. These new strains were then diffused to farmers' fields through a national demonstration program that not only confirmed research results, but also ensured that agricultural workers became familiar with the project and the economic opportunities it offered. The impact was immediate: wheat production rose to 17 million tons in 1968, an increase of 5 million tons over the record harvest of 1964. Such success was later replicated with rice, then cotton, sugarcane, millet, and oilseed. The impact of the green revolution on poverty reduction was, however, limited by the fact that it was not accompanied by widespread land redistribution.

Source: World Bank 1999.

Knowledge cures. The production, dissemination, and application of knowledge in the health arena have had tremendous implications for the well-being of individuals. One of the most successful examples is the treatment of river blindness, which afflicts much of Sub-Saharan Africa, from Senegal and Ethiopia in the north down to Angola and Malawi. Over the last 30 years, a broad international partnership has defeated the disease in many parts of West Africa and is making rapid progress in the remaining areas of western, central, and eastern Africa where the disease is endemic. Before control programs began, tens of millions were infected and hundreds of thousands suffered from total blindness in 30 countries. From the program's modest beginnings in 1996, it is estimated that, in 2007, it will reach 65 million people annually (World Bank 2004f).

Knowledge protects. The December 2004 tsunami that struck coastlines on the Indian Ocean, killing hundreds of thousands of people, showed the importance of knowledge and information dissemination in mitigating the effects of natural catastrophes. Effective satellite and early-alert radio systems—such as those monitoring the Pacific Rim—could have prevented many deaths. It has been decided to establish such a system by 2010 (UNESCO 2005). Another good example of how knowledge can be put to work is the RISEPAK Web site initiative, a joint effort of the World Bank, Harvard University, Lahore University of Management Sciences, World Online, and Pomona College. RISEPAK was created to help relief efforts penetrate remote areas in the wake of the devastating earthquake that shook Pakistan in October 2005, damaging 4,000 villages—2,500 severely. The RISEPAK Web site disseminated information on village damage, demographics, access, and relief. It also fed field information into a central, public database, improving the targeting and coordination of relief work.

Knowledge as the Enabler of Institutions and Governance

It is well established that investments in education give rise to a wide range of non-economic benefits. The creation of economically relevant knowledge, skills, and competencies has benefits beyond work performance in the areas of public health, crime fighting, parenting, community participation, and social cohesion (OECD 1998).

There is an obvious correlation between education levels and the occurrence of civil strife and ethnic wars. In addition, embedded knowledge is of fundamental importance in helping societies structure or restructure themselves, as during post-conflict rebuilding. For instance, calling upon traditional chiefs to solve land issues was essential to the rebuilding of the Mozambican economy following the civil war of the 1980s and 1990s (Mamade, Ahmend, and Easton 2003).

Knowledge is also crucial in the policy-making process. In fact, most development policies are based on the identification and dissemination of good policy practices to all aspects of public administration. Such knowledge can be acquired through dissemination activities, study tours, experiments, and so on. Knowledge and the application of what has worked elsewhere serve as the basis of pragmatic approaches to development strategies and policies. This must be complemented by a deep knowledge of the idiosyncrasies of the society in which policy measures will be implemented.

The channels by which knowledge promotes development are more advanced and multiply faster in societies that have mechanisms in place to enable the transfer of ideas from one time period to another (political stability), from one location to another (regional equalization), and across different groups (inclusion).

How Knowledge Reshapes Economies

Societies have clearly been entering a new postindustrial era over the last few decades. Rapid progress in microelectronics is leading to generalized automation. The Internet has changed every aspect of communication. The role of manufacturing in economies has diminished, while that of services has increased. Scientific advances are dramatically changing how we put life, matter, and energy to use. Extraordinary advances in telecommunications, which sustain the globalization process, have enabled truly worldwide competition. Together with new transport technologies, such advances are shrinking time and distance.

New Global Competition: Trade and Technology

Recent telecommunications advances have accelerated globalization trends. Whereas trade represented 28 percent of world GDP in 1970, it represented 47 percent in 2000. And the share of R&D-intensive production in trade flows has also increased steadily (figure 1.4).

Multinational companies (MNCs) are at the center of evolving cross-border patterns of trade, foreign investment, and technology and knowledge transfers. Such

Figure 1.4 *Growth in World Trade in Manufactures, by Level of Technology of Traded Goods, 1994–2003*

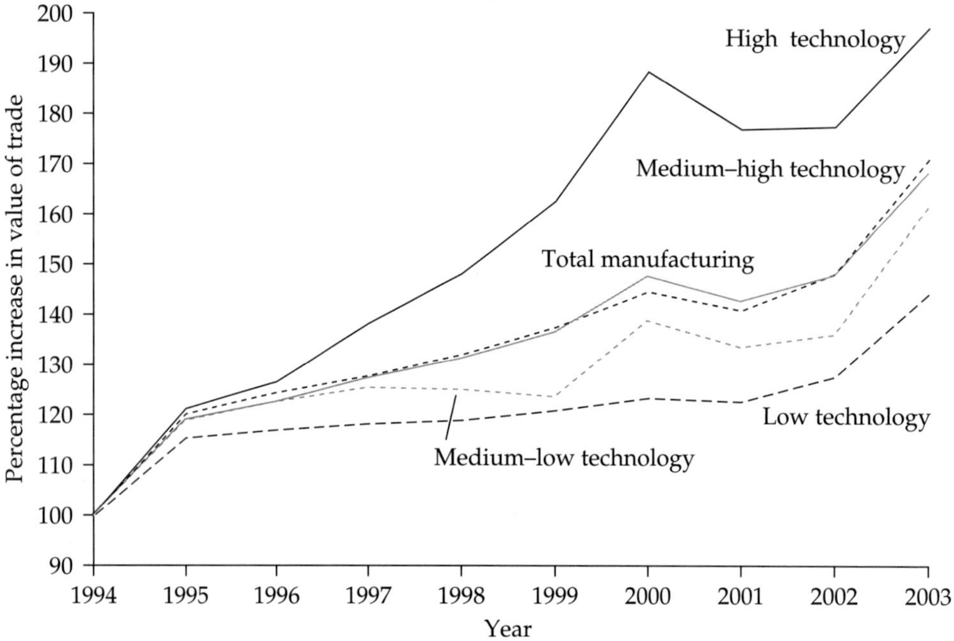

Source: Authors' calculations from World Bank internal data.

firms undertake a large and growing share of world trade; an estimated 30 to 40 percent of international commerce is conducted within them. MNCs also account for the bulk of cross-border investment activity and are leading disseminators of innovation.

New GDP Composition: The Services Sector

In parallel to productivity gains and the automation of manufacturing activities, the relative importance of the service sector has increased.

In advanced countries, the services field now occupies some 70 percent of the labor force (figure 1.5). Services requiring higher qualifications (such as financial services and telecommunications) are growing the fastest. This upward trend in the importance and sophistication of services is also apparent in low- and middle-income countries, and its effects reach far beyond the service sector. Services are used intensively in the production of all goods; they account for an estimated 10 to 20 percent—and more in some cases—of production costs in both industry and agriculture.

The Newly Dominant Position of Human Capital

In line with these changes, tertiary education has become far more widespread than in earlier generations (figure 1.6). The development of a class of knowledge workers (Drucker 1996) is transforming a labor force that is increasingly called upon to think in symbols. This is evident across sectors—among software designers, nurses who monitor hospital patients, farmers who manage automated stables, and so on.

Figure 1.5 The Growing Contribution of Services to GDP

Percentage of total employment, 1990–2003

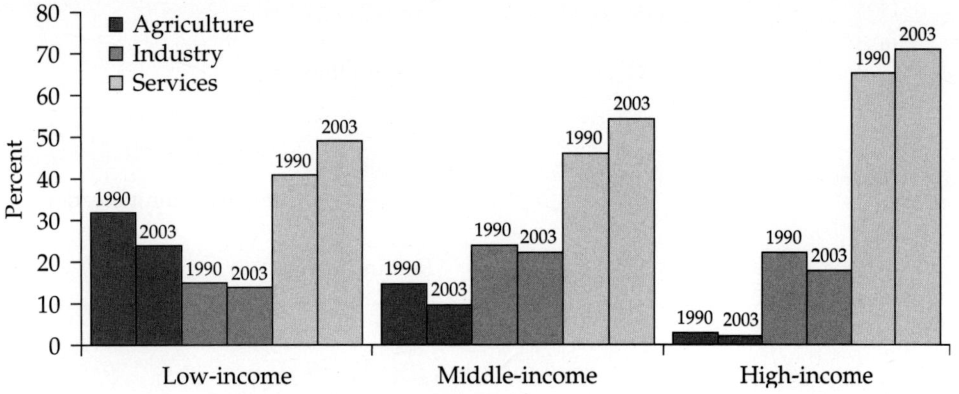

Source: Authors' calculations from World Bank internal data.

Figure 1.6 The Rise of Human Capital

Percentage of population with tertiary-level education, 2002

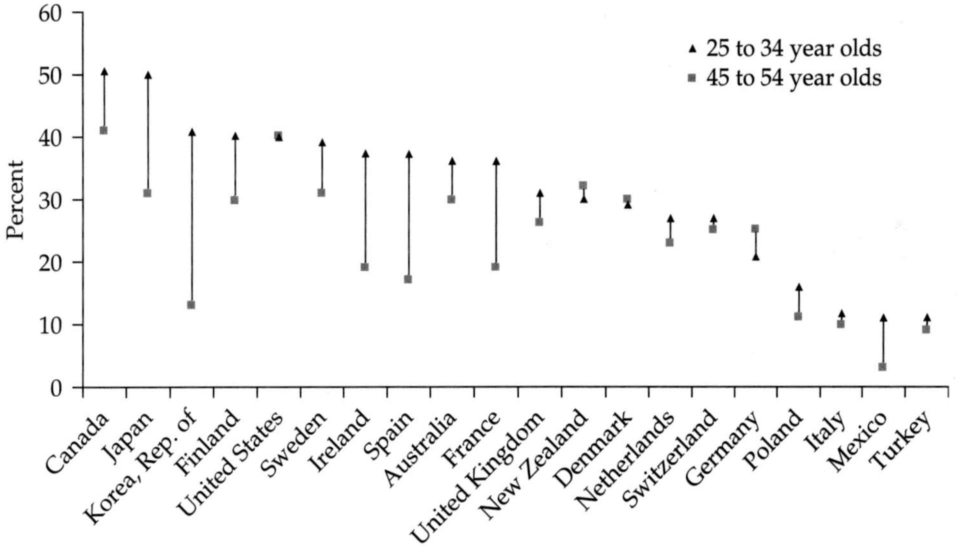

Source: OECD 2004a.

As electronic devices and robots accomplish more and more human tasks, the need for routine skills—both manual and cognitive—is decreasing, while the demand for expertise and communication skills of all kinds is rapidly rising (see figure 1.7 for the example of the United States). The need to understand and interact with images and data is increasing the demand for analytical and integrative skills in even basic functions. This phenomenon stretches educational and organizational systems, as they straddle specialized and generalized skills.

Also, although research on the impact of the knowledge economy on employment creation and growth is still at a preliminary stage, efforts have begun,

Figure 1.7 Changes in Job Tasks and Skill Demands in the United States, 1960–98

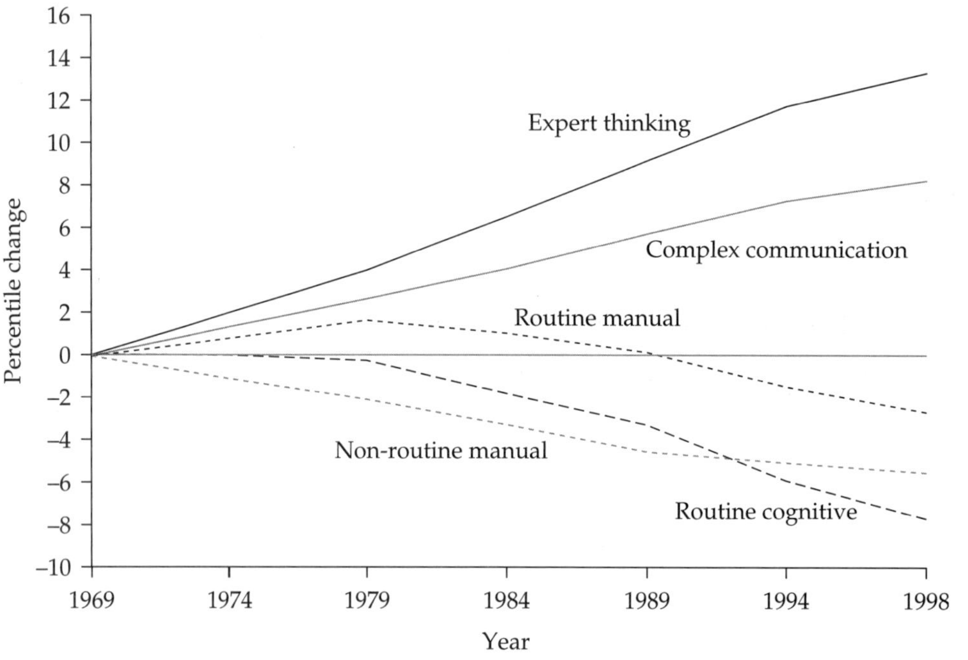

Source: Autor, Levy, and Murnane 2003.

particularly in advanced economies, to quantify its effect. Brinkley and Lee (2006) studied rates of job creation in knowledge-based sectors and low-knowledge sectors in the European Union and United States over a 10-year period (table 1.2).[5] They found that knowledge-based industries created twice as many new jobs in the United States and four times as many in Europe.

A New Investment Allocation

In most economically advanced countries, investments in intangibles—defined as including R&D, software, and training—are now more important than investments in capital goods (figure 1.8). Investments in knowledge as a percentage of GDP among Organisation for Economic Co-operation and Development (OECD) countries increased by almost 1 percent over the period 1994–2002, while investment in equipment and machinery diminished by 0.5 percent.

Innovation processes are changing significantly: science and industry are more closely integrated, and the gestation period for new products is becoming much shorter. Also, the globalization of R&D is becoming a reality; half or more of all applications to U.S. and European patent offices are of foreign origin, and some 14

5. The Eurostat definition of knowledge-based industries is based on four-digit SIC codes used by the OECD. It covers employment in, among other areas, telecommunications, software and research, and certain public sector activities including health, social work, and education. Less knowledge-based industries and services include medium- and low-tech manufacturing, construction, agriculture, energy and water, retail, hospitality, and all other services.

Table 1.2 *Job Creation by Knowledge Content of Sector in the United States and Europe, 1995–2005*

Sector	*Change in number of jobs, 1995–2005* U.S.	EU15
High–medium tech manufacturing	–15.7%	–2.4%
Knowledge-based services	+27.2%	+30.7%
All knowledge-based activity	+20.9%	+23.9%
Low–medium tech manufacturing	–18.3%	–7.5%
Less knowledge-based services	+12.7%	+13.5%
All non-knowledge-based activity	+10.2%	+5.7%
Total employment	+14.0%	+12.6%

Source: Brinkley and Lee 2006.

Figure 1.8 *Investment in Knowledge as a Share of GDP, 2002*

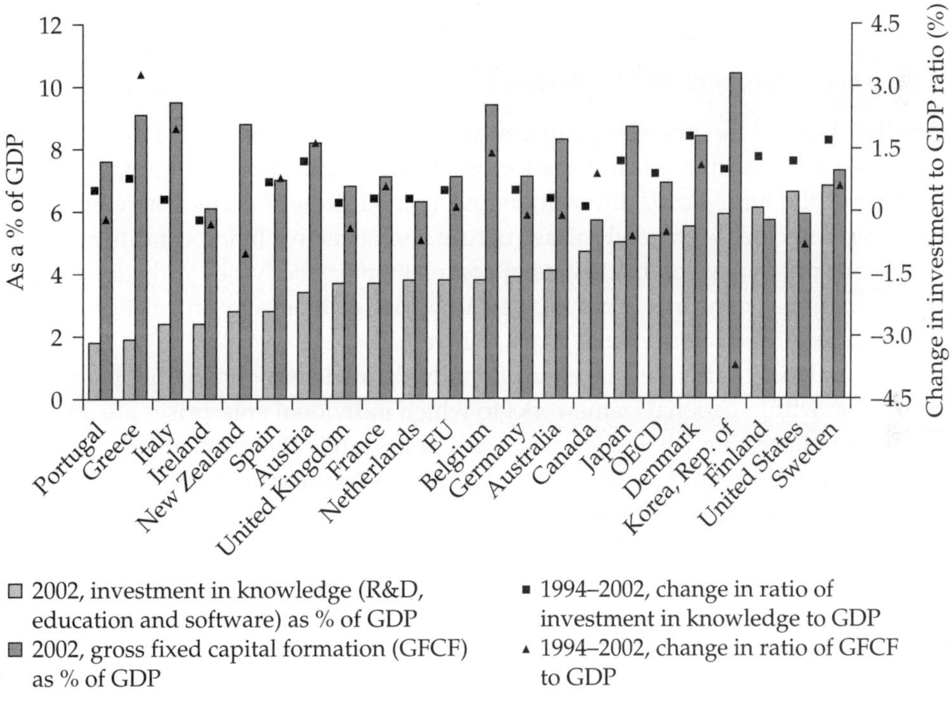

Source: OECD 2005.

Note: Date ranges are 1994–2001 for Greece and Italy, 1995–2002 for Korea. Data on investment in knowledge and GFCF in EU and OECD exclude Greece and Italy. Data on investment in knowledge and GFCF in Greece and Italy are from 2001. Data on change over time in EU exclude Belgium, Greece, and Italy. Data on change over time in OECD exclude Belgium, Greece, Italy, and New Zealand.

percent of all domestic patent applications are owned or co-owned by a foreign resident (OECD 2006b).

Intellectual assets now contribute more to the market value of firms. In 2002, such assets were estimated to represent more than 70 percent of market value on the U.S. stock exchange, up from less than 40 percent in 1982 (figure 1.9). The decrease between 1997 and 2002 can be attributed to the dot-com bubble burst of 2000–01 (because high-risk stocks generally include more intellectual assets).

Figure 1.9 Contribution of Intellectual Assets to U.S. Market Value, 1982–2002

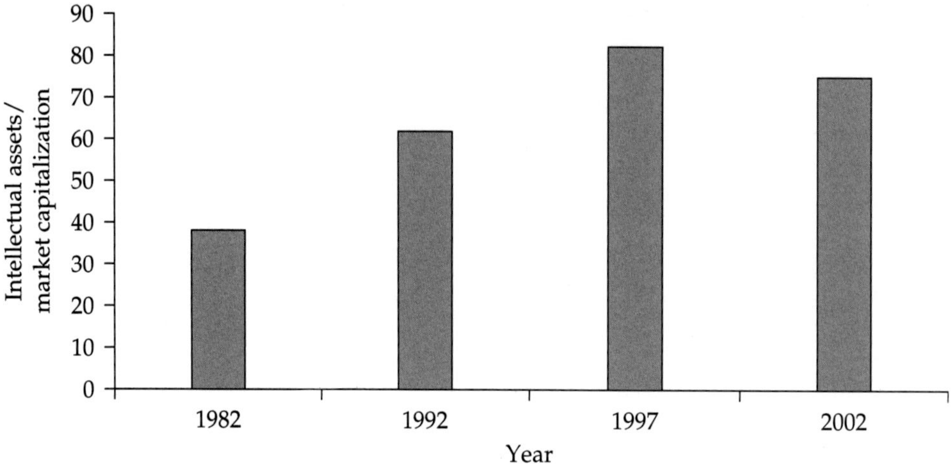

Source: Kaplan and Norton 2004.

A New Local-National-Global Pattern

Another facet of the knowledge-driven transformation of economic development processes is the increased importance of the local dimension, epitomized by the coining of the term *glocal.*[6] Innovation and related growth occur where entrepreneurship coincides with good infrastructure and attractive living conditions. Similarly, enterprises develop in clusters of complementary activities with connections to international networks. This has led to a loss of hierarchy in relations at the national, local, and international levels (figure 1.10).

International trade rules have traditionally defined sector competition and specialization within national frameworks to which individual enterprises are attached and which determine wholesale costs, retail prices, and other economic parameters. In the emerging pattern, interactions between global and national entities determine the rules of the game, while interactions between global and local entities form the basis of market interactions, and national-local relations determine community resources. Firms operate in clusters and largely through international networks (of suppliers, customers, scientific and technical researchers, and so on—see chapter 4).

Challenges and Opportunities

It is for all these reasons that the term *knowledge economy* (KE) has been coined. Its meaning is broader than that of *high technology* or the *new economy,* which are closely linked to the Internet, and even broader than the often-used *information society.* Its foundations are the creation, dissemination, and use of knowledge. A knowledge economy is one in which knowledge assets are deliberately accorded more importance than capital and labor assets, and where the quantity and sophistication of the knowledge pervading economic and societal activities reaches very high levels.

6. *Glocal* is a term combining "global" and "local." Though it originated from Japanese business practices in the 1980s, it was first popularized in the English-speaking world by the British sociologist Roland Robertson in the 1990s.

Figure 1.10 The New Competitive Arena

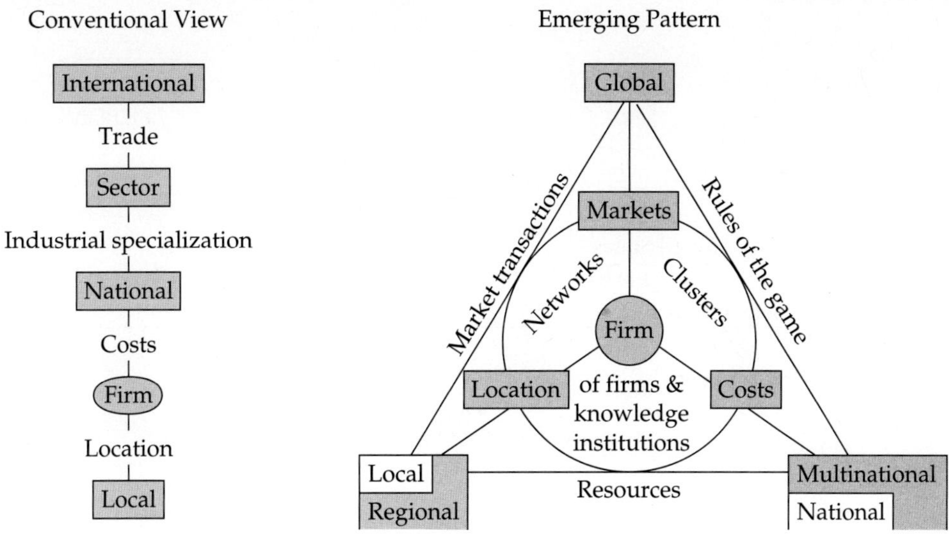

Source: OECD.

Coping with Knowledge-Based Economic Competition

Industrialized countries, for which the term *KE* was initially forged (OECD 1996), are coping unevenly with the new realities. The nations of North America seem to have benefited quickly from the new opportunities offered, with a higher growth rate and higher productivity performances over the last 15 years or so. Gaps in income per inhabitant between North America and Europe have increased. In Europe, small, dynamic economies such as Finland and Ireland have become models of knowledge-based growth and competitiveness, while larger continental economies such as France and Germany—which led the technological and industrial race in past decades—have had difficulty adjusting.[7] Meanwhile, Japan has experienced a difficult decade, with slow growth caused by a variety of factors, but has continued to build KE assets (by increasing spending on basic research, for example) and to maintain the competitive edge of its global manufacturing companies. The Republic of Korea, meanwhile, has actively embarked on a KE track to recover from the financial crisis of 1997–98.

There is a strong correlation between innovation performance, total factor productivity, and economic growth in OECD countries (figure 1.11). Nordic and English-speaking countries have, as a whole, performed better than others.

The transition economies of Eastern Europe have had difficulty coping with the new knowledge-based competition, although they benefited from considerable past investments in education and science. Smaller economies such as Hungary, Slovenia, and Estonia have coped well and taken advantage of European enlargement. Estonia, in particular, has adopted an aggressive KE approach. However, a number of other new EU members and candidates are undergoing a more painful adjustment process. The Russian Federation and other countries of the former Soviet

7. The European Union has launched an explicit KE strategy (the Lisbon Agenda) with the aim of becoming the world's most competitive economy in 2010.

Figure 1.11 *Uneven Innovation Underlying Fluctuations in Total Factor Productivity and Economic Growth*

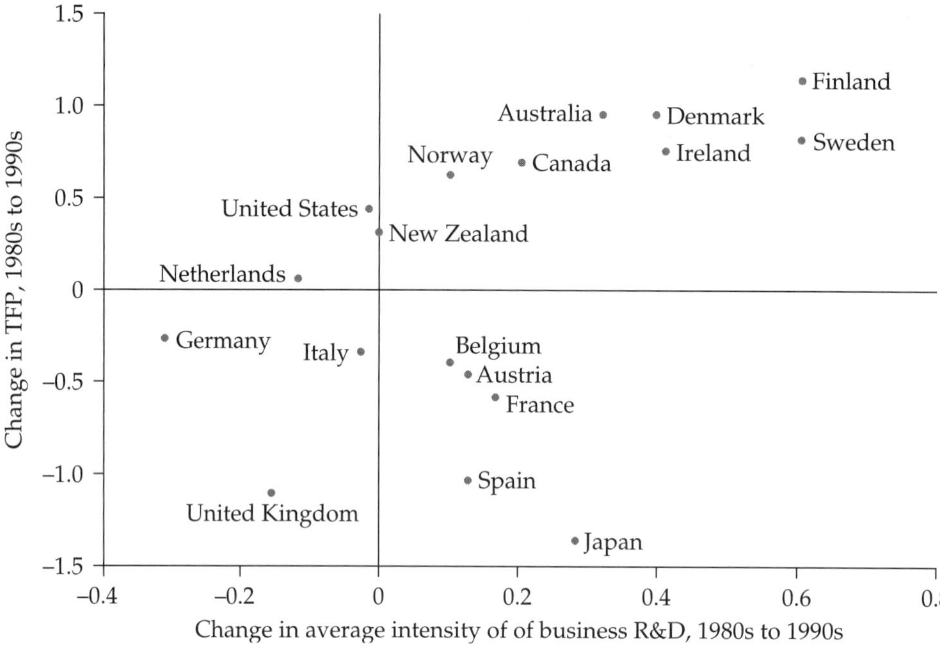

Source: OECD.

Union have yet to demonstrate their capacity to make use of a knowledge potential that was considerable at the time when the Berlin wall fell but eroded rapidly owing to the emigration of highly educated people.

Among medium- and low-income countries, Chile, Malaysia, and Tunisia have clearly taken a knowledge-based approach to increasing competitiveness and growth. According to a recent World Bank study on economic growth in the 1990s (2006d), countries with successful growth—defined as those that both caught up with advanced countries and sustained growth over time—did so by combining three important factors: capital accumulation, efficient resource allocation, and technological catch-up. The 18 successful countries were China, Vietnam, Republic of Korea, Chile, Mauritius, Malaysia, Lao People's Democratic Republic, India, Thailand, Bhutan, Sri Lanka, Bangladesh, Tunisia, Botswana, Indonesia, Arab Republic of Egypt, Nepal, and Lesotho. The report underscores the importance of technological catch-up and its translation into economic growth through increases in total factor productivity, which accounted for between one-half and three-quarters of economic growth in all countries listed.

The report also confirms that productivity gains should be considered in a wide sense—not only in terms of technological change, but also including institutional innovations, which are just as important for productivity as breakthroughs in science and technology. Such gains are also stimulated by internal competition, openness to external markets, and the role of foreign direct investment (FDI) in particular.

Each government among the 18 countries listed played a unique role in the growth process. China embarked on a knowledge-based growth track by attracting massive FDI and then building an indigenous knowledge base through huge investments in education and research. India has succeeded by making the best use of its elite institutions and exploiting international IT-related opportunities, in part through the deft use of knowledge assets.

More generally, as will be demonstrated in chapter 2, there is a distinct KE model and process for countries at all levels of development. Before describing the model further, it is useful to discuss how the knowledge revolution and its implications have changed the prospects of the developing world.

Threats and Opportunities for Developing Countries

Globalization and the knowledge revolution present both challenges and opportunities to developing countries. On the one hand, there is the threat of a widening in the existing knowledge gap with industrialized countries. Indeed, research and innovation capabilities—measured by the usual indicators of R&D investments (expenditures, researchers) and outputs (scientific articles, patents)—tend to be more concentrated in industrialized countries.

On the other hand, the digital gap—differences in telephone and Internet use—is being gradually reduced (figures 1.12 and 1.13), although this does not reflect the considerable inequalities in Internet access among the poor and the rich in developing countries, or the mediocre quality of Internet infrastructures (in terms of bandwidth and so on).

For developing countries, easier access to global knowledge and technology is crucial. Relevant knowledge and modern technology can be decisive in helping such countries reach several of the Millennium Development Goals[8] at a very low cost.

Figure 1.12 *Mobile Telephone Subscribers per 100 Inhabitants, by Region, 1994–2004*

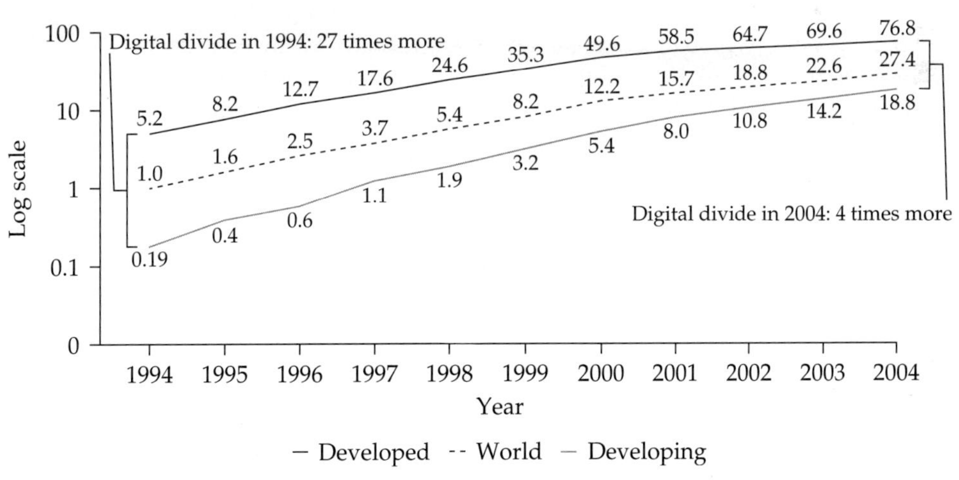

Source: International Telecommunications Union, http://www.itu.int/ITU-D/ict/statistics/ict/index.html and http://www.itu.int/ITU-D/ict/statistics/ict/graphs/mobile.jpg

8. The eight Millennium Development Goals are to (1) eradicate extreme poverty and hunger, (2) achieve universal primary education, (3) promote gender equality and empower women, (4) reduce child mortality, (5) improve maternal health, (6) combat HIV/AIDS malaria and other diseases, (7) ensure environmental sustainability, and (8) develop a global partnership for development.

Figure 1.13 *Average Annual Growth of Internet Users per Capita*

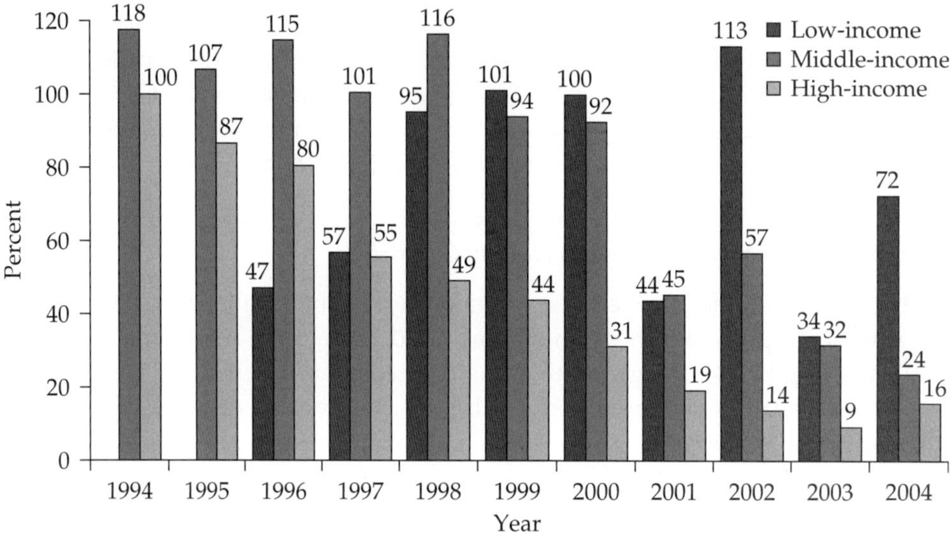

Source: Authors' calculations from World Bank internal data.

The role played by mobile phones in the poorest parts of Africa illustrates this point. The number of subscribers has grown rapidly. This has had a positive impact on the economy in various ways. It enables fishermen and farmers to check prices in different markets before selling their products, makes it easier for people to look for jobs, and reduces travel costs and time. It also lowers transaction costs and broadens trade networks. A research study has found that, in a typical developing country, an increase of 10 mobile phones per 100 people boosts GDP growth by 0.6 percent.[9]

More generally, globalization can be a boon for developing countries that are able to provide their populations with at least minimal exposure to modern technology and a modicum of technical culture while creating an environment that is attractive to foreign and domestic investors. Such countries can take advantage of their relatively low labor costs, as China and India have done—the first to become the world's factory and the second the world's software house. In a similar vein, a number of developing countries are exploiting their geographic and cultural assets for tourism.

Globalization trends have been accompanied by the increased migration of highly qualified labor from the developing world to industrialized countries where their qualifications are in high demand. This "brain drain" can also become a source of knowledge in the long term, whether through expatriates' links with their countries of origin or their eventual return. Taiwan (China), Israel, India, and Ireland illustrate the assets that emigration provides (see chapter 8).

Nonetheless, much is needed to become a vibrant knowledge economy—often more than what was needed to succeed among traditional economies. Then, competition was a matter of capital investments in natural resources or low-cost, unskilled labor. Now, facing world competition means climbing up the value chain. And success in the climb means upgrading the labor force and ensuring efficient

9. "Calling Across the Divide," *The Economist,* March 10, 2005.

Figure 1.14 *Typical Division of Labor in Global Production Networks*

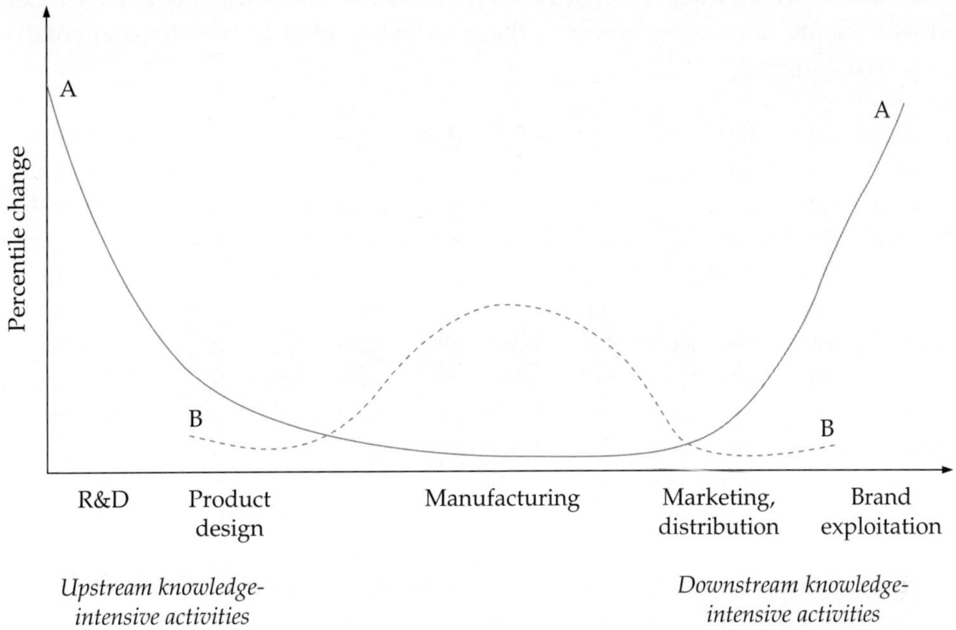

AA: Typical value-added curve for technological leaders in control of global production networks

BB: Typical value-added curve for low-wage participants in foreign-controlled global production networks

Source: World Bank 2003g.

telecommunications and logistics. A knowledge economy requires a significant segment of highly educated people, not simply a population with a basic education.

While low labor costs alone can attract FDI and boost economic growth, on their own they also present the risk of trapping economies in the manufacturing part of the production process (figure 1.14).

Global Issues

The number of major challenges facing the world's economies is mounting, in part because of globalization and the recent technological revolution. Among these challenges are growing fragility in the world community, widening global economic imbalances (all the more difficult to reduce as China and India become major economic players), unsustainable urbanization, and increasingly evident environmental and resource constraints on economic growth.

Knowledge and innovation can help nations face these challenges, several of which are outlined below.

FRAGILITY. Various factors make the world community more fragile, with greater risks of systemic propagation effects and paralysis. These include uncontrolled epidemics such as bird flu, global financial speculation in interconnected markets, terrorist attacks on sensitive points (such as major trade or oil routes), proliferation of weapons of mass destruction, and so on. Such risks result, in part, from the increased integration of economies and societies, which ICTs have accelerated. At the same time, however, these technologies help monitor and control potential dangers.

IMBALANCES. Economic globalization has been accompanied by a redistribution of production through offshoring and outsourcing. FDI has tended to concentrate in a few regions, primarily China and Eastern Europe (following the fall of the Berlin wall). For lower-skill industries, this has led to drastic and permanent employment shifts worldwide. High-income countries have lost jobs, and low- to medium-income countries have lost export and employment opportunities. This trend will likely increase in the coming years and continue to affect service industries, spurred by the rapid growth of India. Consequences are considerable for regions in great need of employment, such as the Middle East, where it is estimated that some 90 million jobs will have to be created in the next 20 years in order to prevent a further increase in unemployment. At present, 15 percent of the total population—and more than 30 percent of the youth population—is unemployed (Aubert and Reiffers 2003).

UNSUSTAINABLE URBANIZATION. The rapid and anarchic urbanization that accompanies industrialization affects developing countries in particular. In 2003, 48 percent of the world's population lived in urban areas—a 33 percent increase from 1990. It is projected that, by 2020, 4.1 billion people (55 percent of the world's population) will live in urban areas. Almost 94 percent of this increase will occur in developing countries. By 2015, there will be 22 megacities (cities or agglomerations with a population of more than 8 million) and 475 cities with populations exceeding 1 million (Léautier 2006). While urbanization helps renew cultures and brings innovations into people's lives, it is accompanied by a loss of autonomy and by violence, human trafficking, and so on. Coping with urbanization and its side effects is a serious challenge. It requires the capacity to conceive, produce, and disseminate technologies that favor autonomous local development processes. This can help prevent the excessive concentration of populations that can lead to dangerous fragmentation.

ENVIRONMENTAL AND RESOURCE CONSTRAINTS. Finally, it is important to recognize that the rapid emergence of China and India, coupled with global warming, means that the world economy cannot continue to use energy and natural resources at the current rate. Production and consumption systems in both developing and industrialized countries will have to change profoundly. Global innovation is challenged—perhaps to a degree never before experienced—as caps on growth are approached (Rischard 2002).[10]

* * *

---

10. For a list of the major worldwide challenges to be confronted in the coming decades, see Jean-Francois Rischard (2002). Rischard lists 20 key global issues that are classified in three broad categories: global issues; economic and social issues—the size and urgency of which require a global commitment; and those requiring a global regulatory approach. Most issues have important knowledge and technology aspects, whether as identifiable problem sources or as means for solutions.

Knowledge has always played a determining role in the development of societies. In the last two decades, however, a distinct KE model and process have been observable in successful economies worldwide, and among both industrialized and developing countries. Globalization and the fast-moving digital age open new opportunities to developing countries to the extent that those countries follow successful economic models. It is urgent that developing countries proceed with the investments and reforms required to build knowledge-based economies. Chief among those requirements are creating jobs, facing competition from China and India, and meeting environmental challenges.

2

The Knowledge Economy: Analytical Framework and Benchmarking Elements

The knowledge revolution and the technological and economic changes it implies clearly entail the need to rethink countries' overall development strategies. Knowledge- and innovation-related policies should be at the core of those strategies, which, we will argue, should be built on four pillars: the country's education and training base, its information and telecommunications infrastructure, the innovation system, and the overall business and governance framework. The last conditions the effectiveness of investments in the other three.

After describing the four-pillar framework, a framework that will serve throughout the report, we adduce some measures and benchmarks that use the framework to situate countries' progress over time and in comparison to one another. We close the chapter with econometric evidence on the strong causal relationships between the development of the knowledge economy and economic growth.

A Four-Pillar Framework

Rationales

A knowledge economy (KE) relies on knowledge as the key engine of economic growth. It is an economy in which knowledge is acquired, created, disseminated, and applied to enhance economic development. Intuitively, conditions for a knowledge-based development process would seem to include an educated and skilled labor force, a dense and modern information infrastructure, an effective innovation system, and an institutional regime that offers incentives for the efficient creation, dissemination, and use of existing knowledge.

- *The labor force should be composed of educated and skilled workers* who are able to continuously upgrade and adapt their skills to create and use knowledge efficiently. Education and training systems encompass primary and secondary education, vocational training, higher education, and lifelong learning. The weight placed on the different segments will differ somewhat depending on a country's level of development. For example, basic education will receive more attention at low levels of development, as basic literacy and numeracy are necessary foundations on which more advanced skills are built. Similarly, lifelong learning has increasing importance in the current context of the knowledge revolution, which requires constant adaptation of knowledge and know-how. It also grows in importance as the population ages. Globalization, meanwhile, is bridging the distance between basic skill needs and advanced skills, forcing countries to cover a wide educational band even at low levels of development to catch up with advanced economies and then remain competitive.

- A *modern and adequate information infrastructure* will facilitate the effective communication, dissemination, and processin g of information and knowledge. Information and communication technologies (ICTs)—including telephone, television, and radio networks—are the essential infrastructure of the global, information-based economies of our time, as railways, roads, and utilities were in the industrial era. They can considerably reduce transaction costs by providing ready access to information. ICT-related policies cover telecommunications regulation as well as the investments needed to build and exploit ICTs throughout the economy and society through various "e-applications"—e-government, e-business, e-learning, etc. Low-income countries should focus first on the basic ICT infrastructure before promoting advanced technologies and applications.
- An *effective innovation system* is composed of firms, research centers, universities, consultants, and other organizations that keep up with new knowledge and technology, tap into the growing stock of global knowledge, and assimilate and adapt it to local needs. Public support for innovation, science, and technology covers a wide range of infrastructure and institutional functions, from the diffusion of basic technologies to advanced research activities. The former should receive a great deal of attention in developing countries. For most developing countries much of the knowledge and technology that nurtures innovation will originate from foreign sources, entering the country through foreign direct investment (FDI), imports of equipment and other goods, and licensing agreements. Foreign sources are important when the economy is less developed, though imports must not be allowed to obscure or marginalize the country's unique indigenous knowledge assets, such as traditional knowledge. Diffusion of basic technologies should receive a great deal of attention in developing countries.
- The country's *institutional regime,* and the set of economic incentives it creates, should allow for the efficient mobilization and allocation of resources, stimulate entrepreneurship, and induce the creation, dissemination, and efficient use of knowledge. The notion covers a vast array of issues and policy areas, ranging from aspects of the macroeconomic framework, to trade regulations, finance and banking, labor markets, and governance. The latter includes the rule of law and its applications (judicial systems), the quality of the bureaucracy as reflected in measures of government effectiveness, and the level of corruption. Mediocre governance resulting in a poor business climate is the single greatest hindrance to economic and social development in general, and to knowledge-based development in particular (figure 2.1).

The contribution to economic growth of the knowledge pillars sketched out above is supported by ample econometric evidence (box 2.1).

Interactions among the Pillars and Virtuous Development Circles

We have seen that each of the four pillars in the KE framework must function efficiently in order to spur knowledge-driven growth. But more is needed: investments

Figure 2.1 The Economic Dividend of Good Public Governance

Measures of good governance are associated with many desirable outcomes

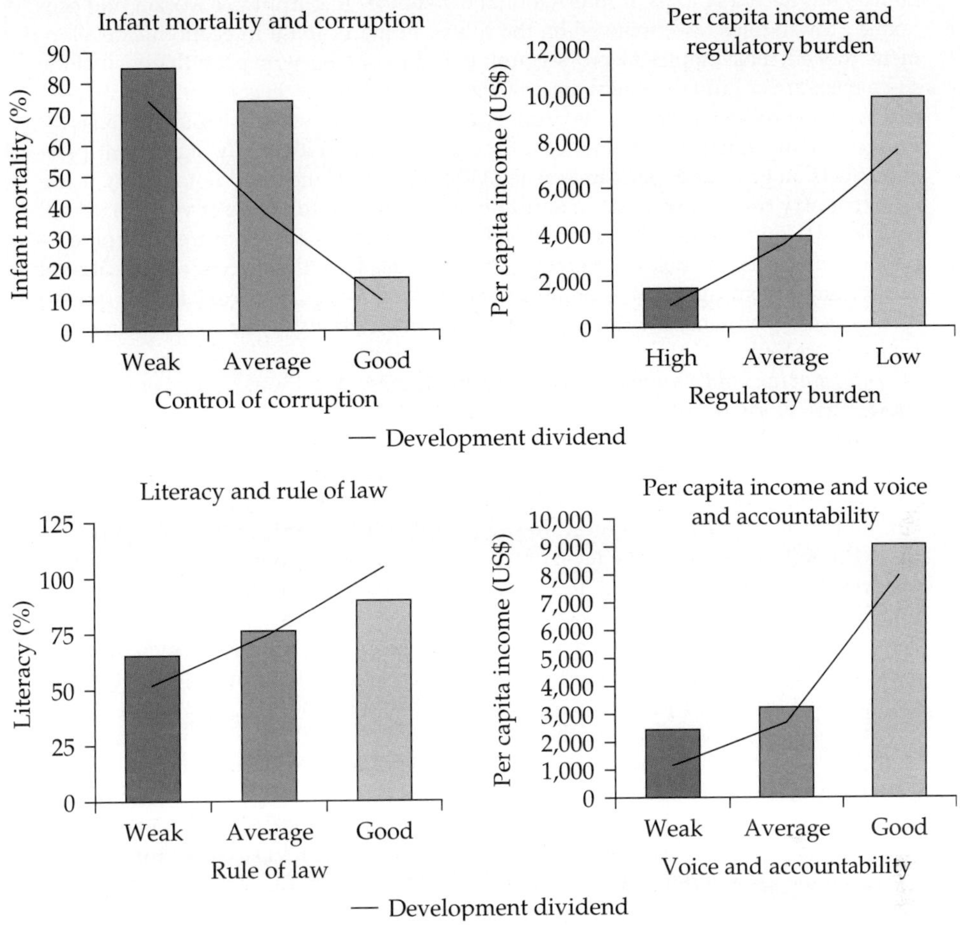

Source: Governance Matters V: Governance Indicators for 1996–2005, accessible at www.worldbank.org/wbi/governance. The Web site of WBI's Governance Program provides a comprehensive analysis of the governance indicators noted above.

in the four pillars must be balanced and coordinated so that the pillars interact to produce benefits greater than those obtainable from their independent operation.

Figure 2.2 illustrates the relationship among the four pillars that make up the knowledge economy. Note that the economic and institutional regime is the base on which the other three pillars are erected. The interdependence of the pillars is illustrated by a few simple relationships. For example, a society must have a minimum level of human capital before it can develop an efficient research and innovation system or reap productivity gains from investments in an up-to-date information technology infrastructure. Likewise, without an effective information infrastructure, innovation systems will find it more difficult to reap the full benefits of global knowledge.

The strong interaction between improvements in the economic and institutional regime and investments in knowledge-related factors (human capital and education, R&D, and ICTs) can be illustrated by the pivotal role played by FDI, which is

Box 2.1 *Knowledge and Growth—Empirical Evidence*

Education. Recent studies of international differences in output per worker and economic growth rates have focused on the role of human capital in economic development. Indeed, most empirical, cross-country studies of long-term growth now include some measure of human capital. Barro (1991), using data for 98 countries for 1960–85 and school enrollment rates at the primary and secondary levels in 1960 as proxies for initial human capital, found that enrollment rates had statistically significant positive effects on growth of per capita real GDP. Similarly, Cohen and Soto (2001), using cross-country time-series data on educational attainment (or average years of school completed), found that education had statistically significant positive effects on economic growth. Hanushek and Kimko (2000) focused on the effects of educational quality on economic growth. Using international test scores as a proxy for the quality of educational systems, they found that educational quality had a positive effect on economic growth.

Information and communications technologies. A growing body of evidence shows that ICTs contribute to a country's *overall* economic growth—and not just growth in its ICT sector. A recent study by the London Business School found that, in a typical developing country, an increase of 10 mobile phones per 100 people boosts GDP growth by 0.6 percentage points (cited in "Calling Across the Divide," *The Economist,* March 10, 2005). Other literature suggests that ICT contributes to labor productivity through increases in the level of ICT investment and growth in the sectors that produce ICTs (Zhen-Wei Qiang, Pitt, and Ayers 2004). While individual firms or sectors of the economy are not automatically made more productive and competitive by ICTs, the potential advantages are numerous. A recent report on ICTs and economic growth in transition economies indicates strongly that ICTs are a major contributor to productivity, profitability, and growth at the level of the firm (InfoDev 2006).

Innovation. Various studies have shown that innovation and the generation of technical knowledge have substantial positive effects on economic growth and growth in productivity. For example, Lederman and Maloney (2003), using regressions on data panels of five-year averages between 1975 and 2000 for 53 countries, found that a 1 percentage point increase in the ratio of total R&D expenditure to GDP increased the growth rate of GDP by 0.78 percentage points. Cincera and van Pottelsberghe (2001) investigated the long-term effects of various types of R&D on multifactor productivity growth, using OECD panel data for the period 1980–98.[a] They found that business, public, and foreign R&D all had statistically significant positive effects on productivity growth. Adams (1990), using numbers of academic scientific papers in various scientific fields as a proxy for the stock of knowledge, found that technical knowledge contributed significantly to the growth in total factor productivity of U.S. manufacturing industries for the period 1953–80.[b]

a. Cincera and van Pottelsberghe (2001) define public R&D as R&D performed by government and higher education sectors, and foreign R&D as business R&D performed in other 15 OECD countries.

b. Adams (1990) used worldwide annual counts of publications in nine sciences: agriculture, biology, chemistry, computer science, engineering, geology, mathematics and statistics, medicine, and physics.

more readily attracted to countries with a good business climate and a high level of education (figure 2.3). Appropriately used, FDI can both improve the business environment and increase knowledge endowments, thus stimulating a virtuous circle of development. But countries must take care not to get caught in a "value-adding curve," wherein low-wage labor is used to add value in global production chains controlled from outside the country (see figure 1.14).

***Figure 2.2** The Four Interactive Pillars of the Knowledge Economy*

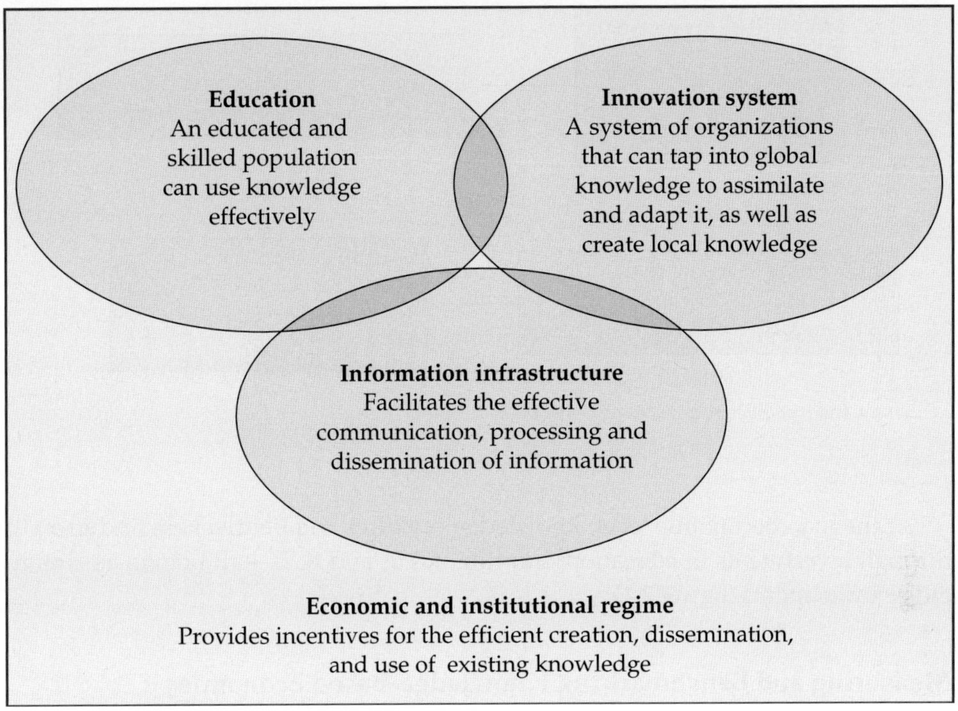

Source: Authors.

***Figure 2.3** Human Capital and FDI Performance*

FDI and investment as a share of GDP are both higher in countries with greater human capital

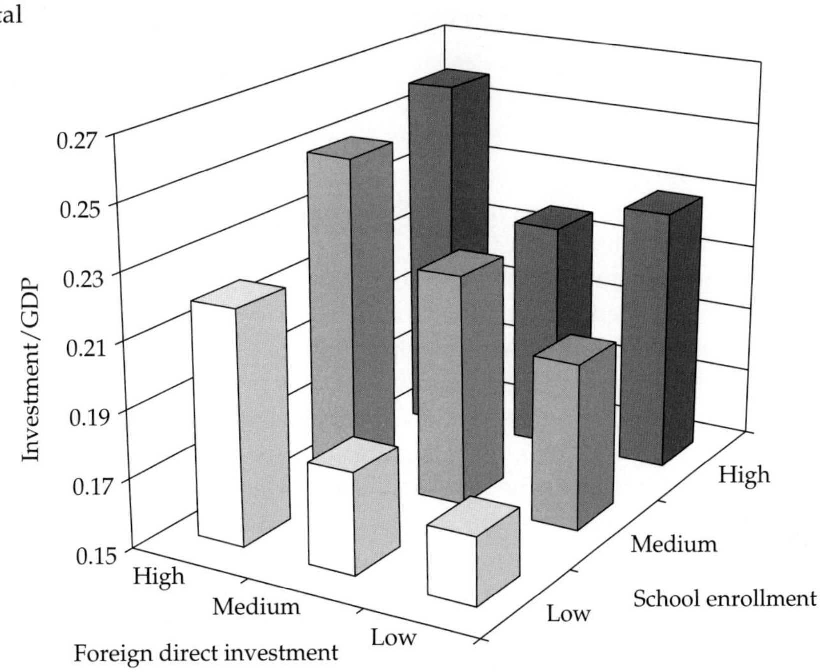

Source: World Bank 2001.

Figure 2.4 *Knowledge-Based Competitiveness and Growth*

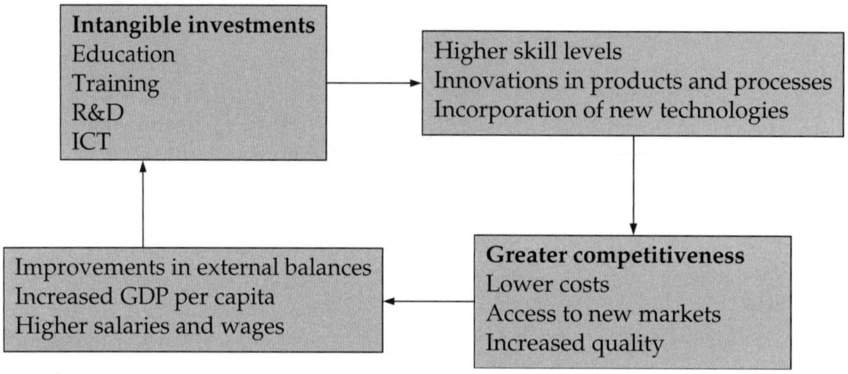

Source: Institut d'Economie Quantitative 2005; authors.

At the macroeconomic level, knowledge nurtures competitiveness and growth through investments in education, training, R&D, and ICTs—all known as "intangible" investments (figure 2.4).

Measuring and Benchmarking Knowledge-Based Economies

Becoming a knowledge economy requires long-term strategies to develop the four KE pillars described above. Countries need to understand their strengths and weaknesses in each domain and then act on that understanding to devise appropriate policies and investments to give direction to their ambitions. Other mechanisms must enable policy makers and leaders to monitor progress against a predefined set of goals.

The Knowledge Assessment Methodology

The Knowledge for Development (K4D) program at the World Bank Institute has developed a database and a set of indexes to measure countries' progress on the four KE pillars. The Knowledge Assessment Methodology (www.worldbank.org/kam) is a Web-based tool that offers a holistic view of the wide spectrum of factors relevant to the knowledge economy. It also provides a basic assessment of countries' and regions' readiness for the knowledge economy (not its performance). Based on the four-pillar framework, it is designed to help countries understand their strengths and weaknesses by comparing their performance over time and their performance relative to other countries. The KAM allows policy makers to pinpoint their country's problems and opportunities, revealing areas where policy attention or investments may be required to make the transition to a knowledge economy.

Comparisons in the KAM are made on the basis of 80 structural and qualitative variables that serve as proxies for the four KE pillars. Because the variables are of different intervals and scales, all are normalized on an ordinal scale from 0 (weakest) to 10 (strongest). Currently, the KAM covers 128 countries and 9 regional groupings. The comparisons are presented on the Web in a variety of charts and figures that highlight similarities and differences across countries. Further technical details on the KAM normalization procedure and data sources are provided in annex 1 to this chapter and on the KAM Web site (www.worldbank.org/wbi/kam).

The KAM "basic scorecard" is a snapshot of the performance of a specific country or region on all four pillars of the knowledge economy. It includes 14 standard variables: 2 performance variables that score the country in terms of GDP growth and its score on the Human Development Index; and 12 knowledge variables used in the Knowledge Economy Index measurement table (table 2.1). The 12 were selected because they are generally available for a long time series and are regularly updated for most countries. More robust data exist for individual countries, but not for enough countries or for long enough periods to make them useful for comparisons.

The three economic and institutional variables aim at characterizing the openness of the economy, the efficiency of government, and the quality of the rule of law. The education variables measure the quantitative achievements of education systems at different levels. The ICT variables measure the penetration of telephones (fixed and mobile), computers, and the Internet. The variables used to measure the innovation pillar are R&D-oriented, but this is due to the difficulty of quantifying innovation, as well as problems in data availability in this domain.

The KAM can be used to compare a country's KE readiness in 1995 and in the most recent period (currently 2004–05). It can also make cross-country comparisons at both points in time.

The basic scorecards of China and India are shown as a spider chart in figure 2.5. The center of the chart denotes the minimum normalized value of 0, and the outer perimeter the maximum value of 10. A larger spider indicates greater readiness for the knowledge economy. The comparison of China and India shows at a glance the relative strengths and weaknesses of the two giants.

***Table 2.1** KAM Basic Scorecard Variables*

Performance
Average annual GDP growth (%)
Human Development Index
Economic incentive and institutional regime
Tariff and nontariff barriers
Regulatory quality
Rule of law
Education and human resources
Adult literacy rate (% age 15 and above)
Secondary enrollment
Tertiary enrollment
Innovation system
Researchers in R &D, per million population
Patent applications granted by the USPTO, per million population
Scientific and technical journal articles, per million population
Information infrastructure
Telephones per 1,000 persons (telephone mainlines + mobile phones)
Computers per 1,000 persons
Internet users per 10,000 persons

Source: KAM version 2006.

Figure 2.5 The KAM Basic Scorecard for China and India

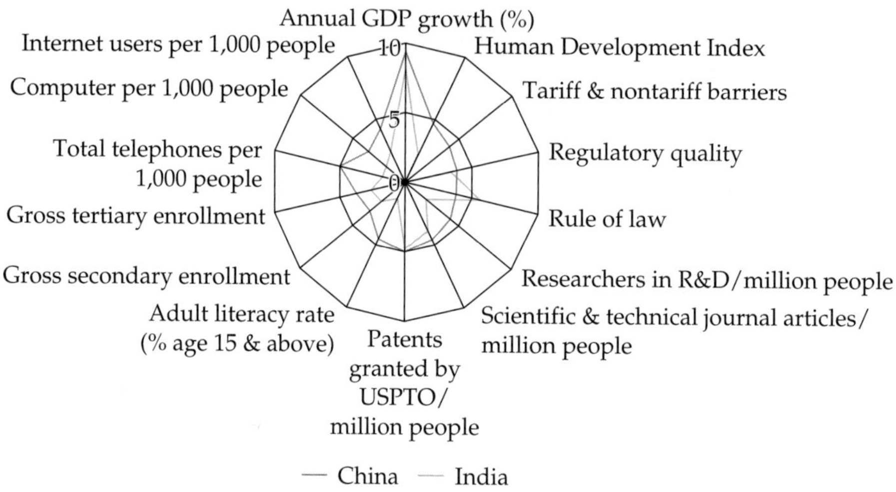

Source: KAM version 2006.

The benchmarking elements in the KAM complement those used by other institutions.[1] The methodology of the United Nations Industrial Development Organization measures countries' capabilities for catching up with other countries. The methodology of the World Economic Forum measures competitiveness assets (box 2.2).

Benchmarking Knowledge Economies: Global Trends

The Knowledge Economy Index (KEI), developed by the World Bank Institute for its Knowledge Assessment Methodology, is an aggregate index that represents the overall development of a country or region in KE terms, summarizing performance over the four KE pillars. It is constructed as the simple average of the normalized values, from 0 (weakest) to 10 (strongest), of the 12 knowledge indicators discussed above with reference to the basic scorecard. Unlike the basic scorecard, however, the KEI does not use the performance variables (see table 2.1). A KEI score close to 10 indicates good development of the four KE pillars compared with other countries; lower scores indicate poorer development.

One can compare countries in terms of their KEI scores in 1995 and 2004–05 (or the most recent year available in KAM).[2] The horizontal axis in figure 2.6 plots the KEI scores of selected countries in 1995, while the vertical axis plots their performance for the most recent year. The diagonal line represents equal values for 1995 and the most recent year. Countries and regions that appear above the diagonal line have improved on the KEI since 1995; countries appearing below the diagonal line have experienced a deterioration.

1. The three methodologies discussed here differ in the rankings they give to countries, although all rank the Nordic countries at or near the top of their lists.
2. KAM users may opt to view the performance of countries or regions on the aggregate KEI or on the individual pillars that make it up.

Box 2.2 *Other Benchmarking Methodologies: WEF and UNIDO*

The World Economic Forum (WEF) measures the factors, policies, and institutions that determine a country's productivity and thus its competitiveness and the level of prosperity it can attain. A more competitive economy is one that is likely to grow faster over the medium to long term. WEF's Growth Competitiveness Index (GCI) is based on three pillars that are critical for growth: the quality of the macroeconomic environment, the state of the country's public institutions, and the level of its technological readiness. The GCI uses a combination of hard data (such as university enrollment rates and the degree of penetration of new technologies) and data from opinion surveys. The survey data, drawn from WEF's Executive Opinion Survey, include concepts such as judicial independence and the extent of inefficient government intervention in an economy. The GCI is currently available for 117 countries (WEF 2005).

The Industrial Development Report of the United Nations Industrial Development Organization (UNIDO 2005) focuses on the capacity of developing countries to catch up with the developed countries. Using data from 135 countries and a mathematical technique known as factor analysis, UNIDO groups and aggregates 29 indicators commonly associated with economic growth into five factors or indexes: (1) knowledge, (2) inward openness, (3) the financial system, (4) the governance system, and (5) the political system. The indicators that make up the knowledge factor include several representing education, innovation, and information and communication technologies, which are very similar to those underlying the Knowledge Economy Index developed by the World Bank Institute for its Knowledge Assessment Methodology. The UNIDO report finds that differences in countries' stocks of knowledge are significantly correlated with differences in income levels across regions. Almost 60 percent of the difference in income between Sub-Saharan Africa and the industrialized countries is attributed to the difference in the stock of knowledge.

The KEI is a *relative* measure. A comparison will help to understand what is being measured. The transition to a knowledge economy is like a global river, with all countries investing significantly in education, research, ICT, and better institutions. But some are investing more than others—or, to maintain the comparison, they are swimming faster. These are the countries above the diagonal line in figure 2.6.

The most developed countries have the highest KE indexes and thus appear in the upper-right area of the chart; low-income countries, with their lower KE indexes, are concentrated in the lower-left area. But some low-income countries have improved their relative position since 1995, reflecting a catch-up trend.

The Nordic economies are among the top performers. Asian economies are slightly behind and have maintained their position since the mid-1990s. Most of the countries of Latin America and the Middle East are middle-income countries. In Latin America, Chile is outperforming others. Brazil has significantly improved its position, thanks to its educational efforts, while Argentina's position has deteriorated owing to the country's weak economic and institutional regime. Among the countries of the Middle East and North Africa, one notes the good performance of several KE pioneers such as Jordan and Tunisia. Among transition economies, the Central European countries that recently entered the European Union distinguish themselves by their good performance, thanks notably to their improved business and governance environment. The position of the Russian Federation has not

Figure 2.6 Relative Performance Over Time of Selected Countries on the Knowledge Economy Index

Improvement or regression on KEI over period from 1995 to most recent period (2004–05 in most cases)

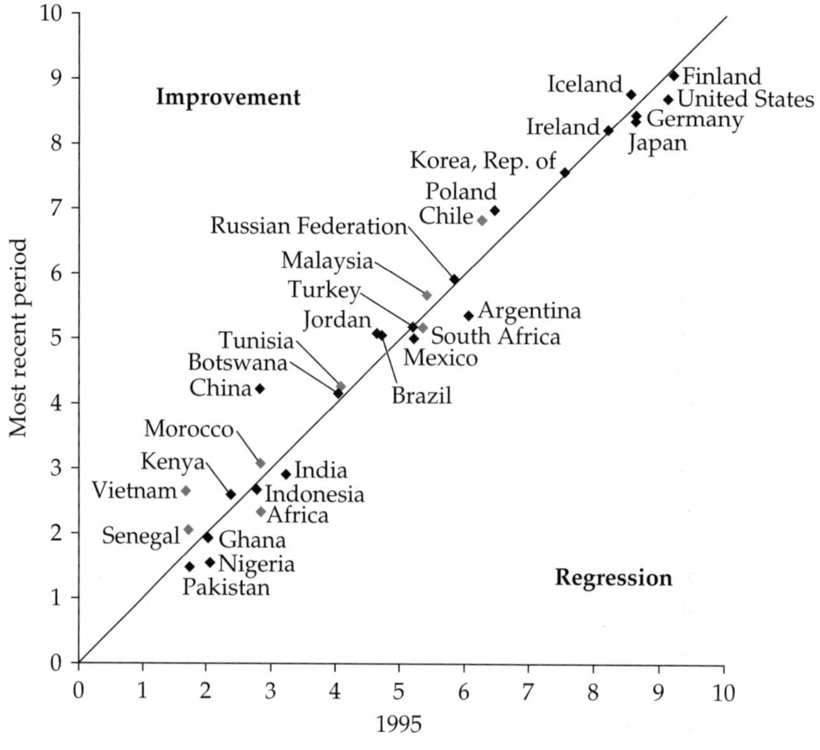

Source: KAM version 2006.

Note: Countries appearing above the 45-degree line have improved their position on the KEI in the most recent period for which data are available relative to their position in 1995 (or the closest available date in the mid-1990s). Countries appearing below the line lost ground during the same period.

changed much since the mid-1990s. China and India seem to have followed contrasting paths, owing to slower reforms in India, although India has made progress on all four pillars.

The Knowledge Economy and Economic Performance

It seems logical that levels of economic development and levels of knowledge should be closely related. The correlation between knowledge accumulation, as measured by the KEI presented above, and levels of economic development is 67 percent (figure 2.7).

The positive correlation between the results of the KEI and the level of economic development does not establish a causal relationship—a high KEI will not necessarily produce a high level of economic development. On the other hand, it is plausible that high-income countries, because they are more affluent, are able to afford greater investments in knowledge and thus score higher on the KEI.

Figure 2.7 The Knowledge Economy and Current Economic Performance

Countries that score higher on the Knowledge Economy Index have higher levels of economic development, and vice versa

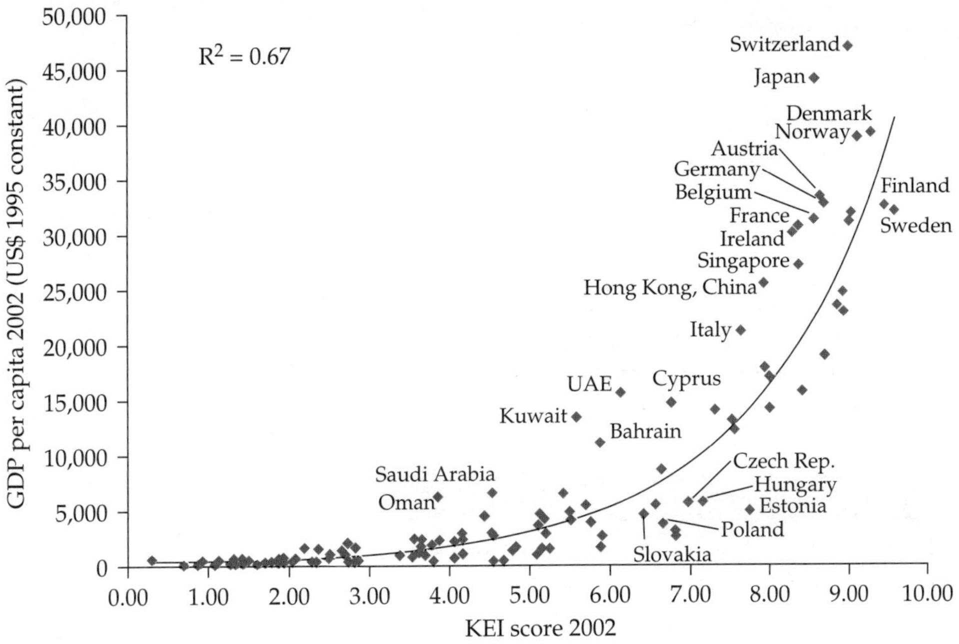

Source: Authors' calculations.

In fact, econometric tests described in annex 2 to this chapter reveal a *statistically significant* causal relationship running from the level of knowledge accumulation as measured by the KEI to future economic growth. The relation between countries' KEI values and the average future rates of growth in output per worker is illustrated in figure 2.8. The figure's horizontal axis plots countries' scores for the KEI for 1995, after accounting for differences in initial real GDP per capita and the growth of capital per worker. The vertical axis plots countries' annual growth in output per worker, averaged over the years 1996–2004 and after accounting for differences in initial real GDP per capita and the growth of capital per worker.

It is clear, therefore, that higher KEI values are associated with higher rates of future economic growth, if other factors are held constant. This suggests that higher levels of knowledge in a society do indeed lead to higher levels of economic growth—and consequently to higher levels of economic development. A one-unit improvement in the KEI, equivalent to moving up one decile or 13 positions in the country rankings, leads to an increase of 0.46 percentage point in economic growth, after accounting for initial conditions, as indicated.

After refining the above analysis by reestimating regressions with the countries stratified into four income categories, we determined that the growth effects of knowledge accumulation are important at all levels of economic development.

These results are important. They confirm that knowledge and its applications have played a major role in the growth of countries. That conclusion justifies plac-

Figure 2.8 *Country Scores on the Knowledge Economy Index in 1995 Correlated with Future Economic Growth*

KEI values rise with average future rates of growth in output per worker

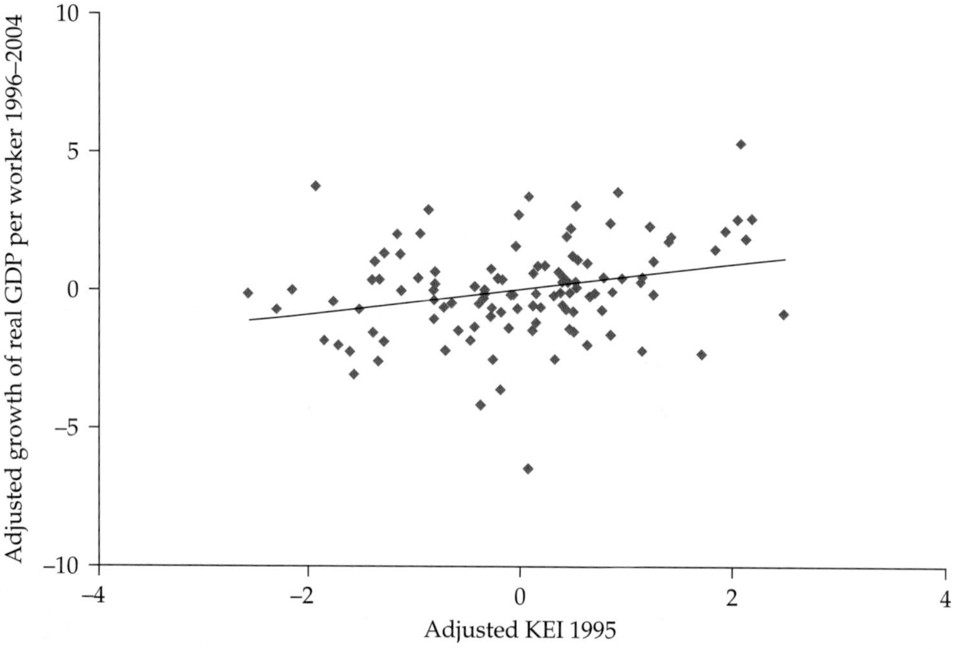

Source: Authors' calculations.

ing knowledge-related policies at the core of today's development strategies. At the same time, it would appear essential to improve the economic fundamentals so as to facilitate knowledge-based growth and accelerate the development process.

Annex 1 The Knowledge Assessment Methodology

Normalization Procedure

The World Bank Institute's Knowledge Assessment Methodology (KAM) consists of data for 132 countries. Included are 81 variables that proxy the four pillars of the knowledge economy (KE), economic and social performance, and gender issues. The normalization procedure used in the KAM is as follows:

- The raw data (*u*) are collected from World Bank data sets and international literature for 81 variables and 132 countries.
- Ranks are allocated to countries according to the absolute values (in raw data) that describe each of the 81 variables (rank *u*). Countries with the same performance are allocated the same rank. Therefore, a country ranked 1 performs the best among the 132 countries in our sample on a particular variable (that is, it has the highest score); a country ranked 2 performs second best, and so on.
- For each country, the number of countries that rank below it (*Nw*) is calculated.
- The following formula is used to normalize the scores for each country, for every variable, according to the country's ranking and in relation to the total number of countries in the sample (*Nc*) with available data:

$$Normalized\ (u) = 10\left(\frac{Nw}{Nc}\right) \qquad (2A1.1)$$

- The above formula allocates a normalized score from 0 to 10 for each of the 132 countries with available data for the 81 variables. The top score is 10, and 0 is the lowest. The top 10 percent of performers get a normalized score between 9 and 10, the second-best 10 percent are allocated normalized scores between 8 and 9, and so on. As mentioned, more than one country may be ranked either at the top or bottom of normalized scores. The 0–10 scale describes the performance of each country on each variable, relative to the performance of the rest of the countries in the sample.

Data Sources

For details about KAM data sources, please visit the KAM Web site (www.worldbank.org/kam).

Basic and Custom Scorecards

The KAM basic scorecard summarizes the performance of a specific country or region on all four KE pillars. It can take a variety of graphic forms—among them spider, diamond, and bar charts. The spider chart is illustrated in the main text of this chapter. Figure 2A1.1 uses a diamond chart to present the basic scorecard of the

Figure 2A1.1 *Basic Scorecard (Diamond Chart): Slovak Republic*

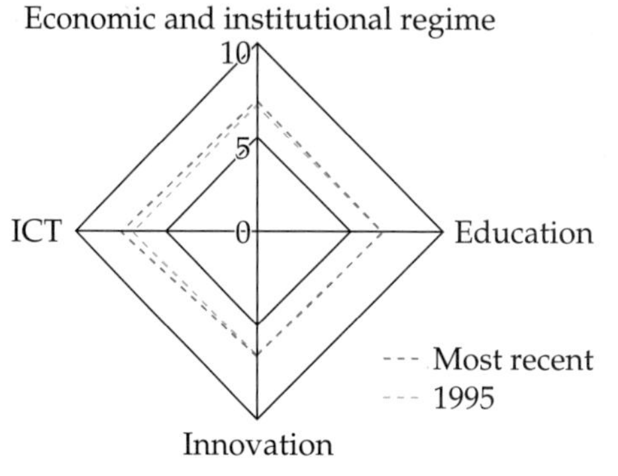

Source: KAM version 2007.

Slovak Republic. Here, only the country's aggregate performance on each of the four KE pillars is shown. The value for each pillar is constructed as the simple average of the normalized values of the three knowledge indicators that proxy each pillar in the basic scorecard. As with the spider chart, a larger or "fuller" diamond chart implies that the country or region is better positioned in terms of KE readiness.

The KAM also allows users to combine variables into custom scorecards for the purpose of making comparisons. One may compare any two or three countries or regions on any of the 81 variables included in the KAM database. Custom scorecards are often used to generate scorecards that focus on a single KE pillar or sector. For example, figure 2A1.2 illustrates the KAM variables for education and training in Estonia and the Europe and Central Asia region, while figure 2A1.3 illustrates the KAM variables for information infrastructure in Finland, the Republic of Korea, and South Africa.

Figure 2A1.2 *Custom Scorecard: The Education Pillar in Estonia and Europe and Central Asia*

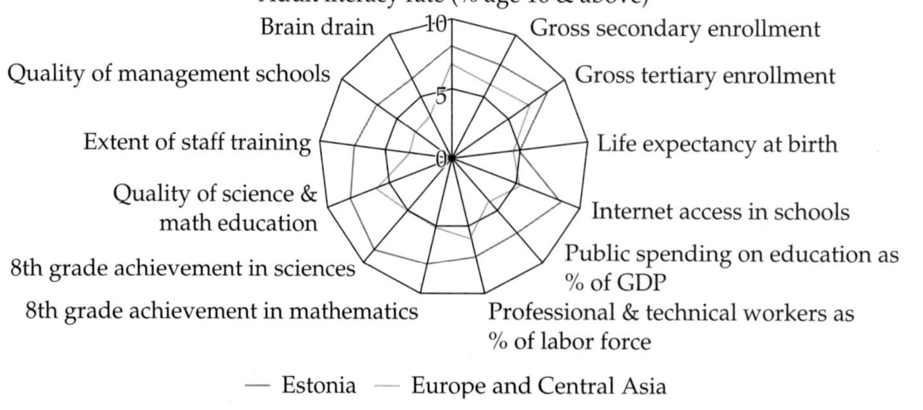

Source: KAM version 2007.

Figure 2A1.3 *Custom Scorecard: The Information Infrastructure Pillar in Finland, Korea, and South Africa*

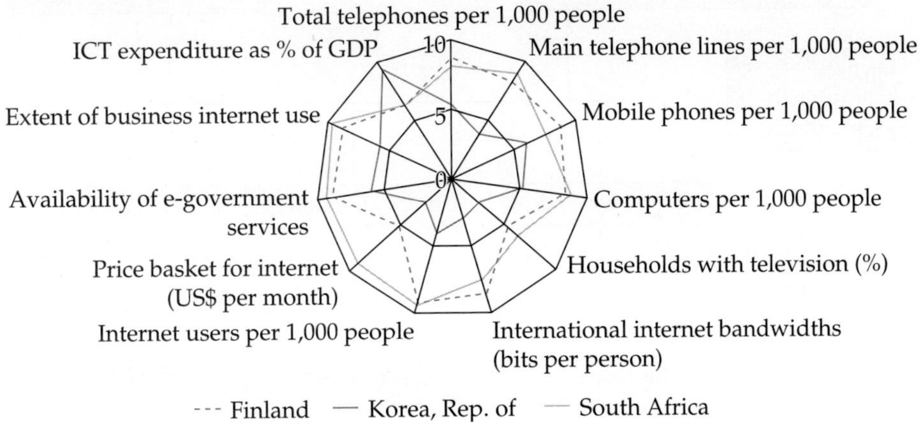

Source: KAM version 2007.

KAM Indexes

Apart from individual indicators, the KAM also produces indexes that summarize country performance on the entire KE framework or portions of it.

The Knowledge Economy Index

The most commonly cited KAM index is the Knowledge Economy Index (KEI), an aggregate index representing the overall level of development of a country or region in KE terms. The KEI summarizes performance over the four KE pillars and is constructed as the average of the values of the four KE pillar indexes.

The Knowledge Index

The KAM Knowledge Index (KI) measures a country's ability to generate, adopt, and diffuse knowledge. This indicates the overall potential for knowledge development in a given country. Methodologically, the KI is the simple average of three KE pillars: education and human resources, the innovation system, and information and communications technology.

Individual Pillar Indexes

The KAM also produces indexes for each of the four KE pillars. The value for each pillar index is constructed as the simple average of the normalized values of the three knowledge indicators that proxy for each pillar in the basic scorecard.

Figure 2A1.4 is a schematic representation of the relationships between the various indexes and the indicators used in the KAM basic scorecard.

Figure 2A1.4 *Relationships of KAM Indexes and Indicators*

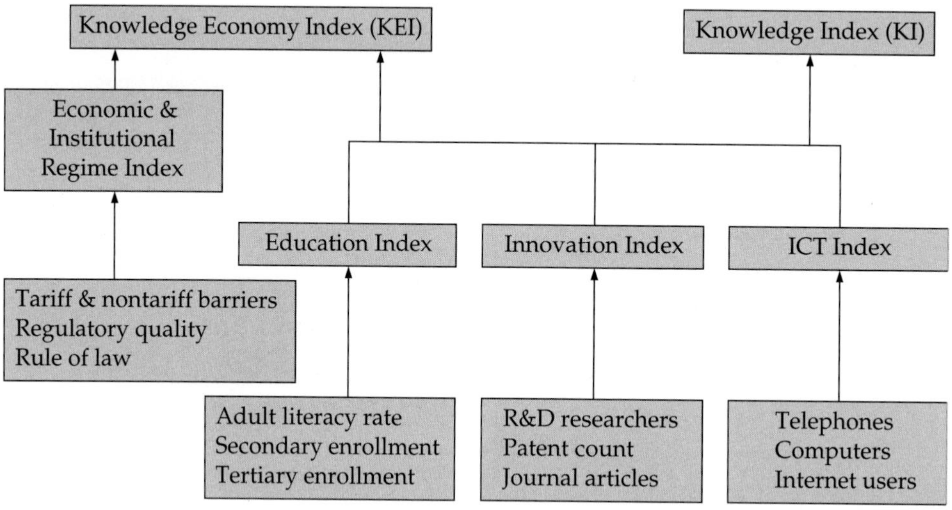

Source: Authors.

Annex 2 The Knowledge Economy and Economic Growth: Econometric Calculations

The Effects of Knowledge Accumulation on Economic Growth

To derive an equation to estimate the effects of knowledge accumulation on economic growth, we use a neoclassical growth model with a Cobb-Douglas specification. We show that output growth per worker depends on (a) capital growth per worker and (b) total factor productivity (TFP) growth, or the current level of technology.

$$\text{growth of output per worker} = \alpha_k\,(\text{growth of capital per worker}) + \text{TFP growth} \qquad (2A2.1)$$

where
α_K is the elasticity of output to capital

Assuming that the key driver of TFP growth is knowledge accumulation, the TFP growth rate can be replaced with the Knowledge Economy Index (KEI), described in annex 1. In addition, in order to avoid reverse causality, we regressed 1995 (past) values of the KEI against output growth per worker during the period 1996–2004. As such, there is little possibility of economic growth rates affecting the KEI,[1] thereby minimizing endogeneity problems.

The literature shows that it is necessary to include a term that captures the initial income of a country. One of the key implications of the neoclassical growth model is that, all else being equal, poorer countries should grow faster than richer ones.[2] To control for initial income, we used real GDP per capita in 1996, which is the beginning of the regression period. In light of the above, the resulting estimating equation is therefore:

$$\begin{aligned}\text{growth of output per worker} = \alpha &+ \beta_1\,[\text{real GDP per capita}_{1996}] \\ &+ \beta_2\,[\text{growth of capital per worker}] \\ &+ \beta_3\,[\text{KEI}_{1995}] \qquad (2A2.2)\end{aligned}$$

1. This would be the case especially if we assumed adaptive, not rational, expectations.
2. For example, if two countries have the same long-run potential level of income, the country with lower current income should grow faster. Lower-income countries tend to have a smaller capital stock (fewer machines, factories, and roads) than higher-income countries. Since capital is relatively scarce, the rate of return on new investments tends to be higher, leading to faster growth. Poor countries also have the advantage of being able to borrow new technologies and tested management practices from richer countries without paying the costs of research and development. Many studies on economic growth have shown that once differences in other important structural and policy variables are taken into account, poor countries do indeed tend to grow faster than rich countries (Barro 1991; Sachs and Warner 1995). This outcome is known as conditional convergence, since the income levels of countries converge over time, conditional on having similar policies, resource endowments, and so on.

***Table 2A2.1** Knowledge and Economic Growth, Dependent Variable: Growth Rate of Real GDP per Worker*

	Reg A1	
Years: 1996–2004	*Estimated coefficient*	*Standard error*
(Log) initial GDP per capita (1996)	−0.9459***	0.3944
Growth of capital per worker	0.3838***	0.0323
KEI 1995	0.4605***	0.1673
Constant	6.6693***	2.6602
R squared	0.5894	
Number of countries	113	

Source: Authors.
Note: Cross-section regression variables were averaged over 1996–2004 period.
*, **, *** denotes significance at the 90, 95, and 99 percent confidence level, respectively.
Robust standard errors obtained using White's correction for heteroscedasticity.

Table 2A2.1 presents the regression results. We see that the estimated coefficient of KEI 1995 is positive and statistically significant.[3] The estimated value of 0.4605 implies that a unit increase in the KEI tends to increase average annual growth of output per worker by 0.46 percentage point. Recall that the KEI ranges from 0 to 10, and a unit increase is equivalent to an improvement of one decile (or about 13 positions) in the ordinal ranking of the 132 countries included in the KAM.

Effects That Hold at All Levels of Economic Development

Knowledge accumulation is important for economic growth at all levels of income. To demonstrate this, we reestimate a variant of equation (2A2.2), this time allowing for different effects of the KEI on economic growth for each of the World Bank's income-level categories: high-income, upper-middle, lower-middle, and low-income in 1995.[4] As a result, the estimating equation becomes:

$$\begin{aligned}\text{growth of output per worker} = \alpha &+ \beta_1\,[\text{real GDP per capita}_{1996}]\\ &+ \beta_2\,[\text{growth of capital per worker}]\\ &+ \beta_3\,[\text{KEI}_{1995}] + \beta_4\,[D_{UM}\times\text{KEI}_{1995}]\\ &+ \beta_5\,[D_{LM}\times\text{KEI}_{1995}] + \beta_6\,[D_L\times\text{KEI}_{1995}] \qquad (2A2.3)\end{aligned}$$

3. Data for real GDP (in constant local currency units), real GDP per capita for 1996 (in constant PPP terms), and labor force figures were taken from the World Bank SIMA database. The capital stock was constructed using gross capital formation (in constant local currency units) obtained from the World Bank SIMA database. The perpetual inventory method was used with an assumed depreciation rate of 5 percent. To calculate the initial value of the capital stock, we used the average growth rate of gross capital formation for the first five years and applied the formula for the sum of an infinite geometric progressive series. For further details, see appendix A of Chen and Dahlman (2004).

where

D_{UM} is the dummy variable for economies in the upper-middle-income category in 1995

D_{LM} is the dummy variable for economies in the lower-middle-income category in 1995

D_L is the dummy variable for economies in the low-income category in 1995

β_3 gives the estimated effect of the KEI on growth of output per worker for high-income economies, which also serves as the benchmark

β_4 gives the estimated difference, in terms of KEI growth effects, between upper-middle-income and high-income economies

β_5 gives the estimated difference, in terms of KEI growth effects, between lower-middle-income and high-income economies

β_6 gives the estimated difference, in terms of KEI growth effects, between low- and high-income economies

In light of the above,

$\beta_3 + \beta_4$ gives the estimated effect of the KEI on growth of output per worker for upper-middle-income economies

$\beta_3 + \beta_5$ gives the estimated effect of the KEI on growth of output per worker for lower-middle-income economies

$\beta_3 + \beta_6$ gives the estimated effect of the KEI on growth of output per worker for low-income economies

Table 2A2.1 presents the results from the estimation of equation (2A2.3). We see that the estimated coefficient denotes that the effect of the KEI on future economic growth for high-income economies is positive and statistically significant. The value of 0.4582 indicates that a unit increase in the KEI in high-income economies in 1995 is associated with a 0.4582 percentage point increase in 1996–2004 in terms of average annual growth of output per worker. Turning to the estimated coefficient of the KEI for upper-middle-income economies, we see that the value is 0.0168 and is not statistically significant. This implies that the estimated effect of the KEI on future economic growth for upper-middle-income economies is larger than that of high-income economies by 0.0168 units, which implies a net effect of 0.4750 units. However, this difference is not statistically distinguishable from zero.

The situation is very similar for that of lower-middle-income economies. In this case, the estimated effect of knowledge on future economic growth is larger than that of high-income economies by 0.0192 units, thereby implying a net effect of 0.4774 units, but this difference is not statistically distinguishable from zero. In the case of low-income economies, the estimated effect of knowledge on future economic growth is smaller than that of high-income economies by a magnitude of 0.0022 units; however, this difference is still not statistically distinguishable from zero. The difference of 0.0022 units implies that, for low-income economies, the estimated effect of knowledge accumulation on future economic growth is 0.4560.

4. Income category classifications were obtained from the World Bank 2005h.

Overall, we see that the effect of knowledge accumulation on future economic growth is positive and does not differ statistically across economies in the different categories of income.

Table 2A2.2 *Knowledge Effects on Economic Growth by Income Category, Dependent Variable: Growth Rate of GDP per Worker*

		Reg A2	
Years : 1996–2004	*Estimated coefficient*	*Standard error*	*Number of countries*
(Log) initial GDP per capita (1996)	−0.9488***	0.4102	113
Growth of capital per worker	0.3836***	0.0393	113
KEI 1995 high-income (benchmark)	0.4582***	0.1617	25
KEI 1995 upper-middle income (difference)	0.0168	0.0641	17
KEI 1995 lower-middle income (difference)	0.0192	0.0939	39
KEI 1995 low-income (difference)	−0.0022	0.3284	32
Constant	6.6641***	2.9749	113
R squared		0.5897	
Total number of countries		113	

Source: Authors.
Note: Cross-section regression -variables were averaged over 1996–2004.
*, **, *** denotes significance at the 90, 95, and 99 percent confidence level, respectively.
Robust standard errors obtained using White's correction for heteroscedasticity.

3

The Process of Building a Knowledge Economy

Drawing on the Lessons of Success

Learning from Others

To understand how to build knowledge-based economies, it is useful to look at countries that have succeeded in setting their growth processes on a knowledge- and innovation-based track—even if the relevant policy actions were part of broader development strategies and an explicit knowledge economy (KE) approach was only recently identified and named. Several cases throughout the world deserve particular attention.

ADVANCED ECONOMIES. Finland is considered by many to be the world's most competitive country. Canada and Australia also enjoy competitive economies. The Republic of Korea and Ireland initiated explicit KE strategies in the past few decades, starting from a low-income base to achieve leading positions in the world economy.

MIDDLE-INCOME ECONOMIES. A few decades ago, the nations of Chile and Costa Rica in Latin America, Malaysia in East Asia, Tunisia in the Middle East, and Mauritius and Botswana in Africa, instituted multisector reforms to attract foreign investment and create a KE-oriented environment.

TRANSITIONAL ECONOMIES. The Baltic countries, notably Estonia, have instituted KE reforms over the past decade that are now paying off.

LOW-INCOME ECONOMIES. Vietnam has developed rapidly by taking advantage of globalization. The African countries of Mauritania, Mozambique, Uganda, and Rwanda are also active in instituting KE reforms (if in a fragmented way) and have enjoyed some economic success.

CHINA AND INDIA. Finally, there are the examples of China and India. These are the two *emerging giants* of our time, and their ascendance has benefited from the selective use of the KE approach.

The experiences of these countries offer answers to the questions of *what to do* to build a knowledge economy and *how to do it*. The examples of the Republic of Korea, Ireland, and Finland are examined in detail in this chapter. Although their economies are now fairly advanced, they offer useful and generally applicable lessons.

Republic of Korea

After World War II, Korea's gross domestic product (GDP) per capita was similar to those of the poorer countries of Asia and Africa. During the four years of privation and destruction endured during the Korean War, conditions became even worse. In 1953, Korea's only assets were a high enrollment in primary schools and an equitable distribution of wealth (WBI and KDI 2007; Dahlman and Andersson 2000). Yet, by early 2006, GDP per capita in the Republic of Korea had grown to be roughly 20 times higher than that of the Democratic People's Republic of Korea and equal to that of the mid-level economies of the European Union (EU).

Korea began its rise by exporting light industrial goods such as textiles and bicycles. In the 1970s, it gradually developed sophisticated production in heavy industry (shipbuilding and automotives), chemicals, and, later, electronics. It obtained needed technology through licensing and original equipment manufacturing (OEM) contracts. Industry developed on the platforms created by large family-owned groups known as *chaebol* and was boosted by demand during the Vietnam War. Big business benefited from targeted government support granted through an approach to industrial policy common in Asia. That approach included industrial regulation, technology upgrades, training, preferential credits, export subsidies, and import restrictions.

Meanwhile, Korea began to develop its research and development (R&D) base through government institutes, initially to facilitate the adaptation of imported technology and later to pursue internal research. The growing industrial sector began to fund R&D. In the last decade, industry funded 80 percent of total R&D costs, which now represent 3 percent of GDP.

Throughout its economic development, Korea has invested heavily in education, starting from a relatively high level of literacy in the 1950s. Its initial priority was basic education, but professional and vocational education gradually received the attention required by the growing industrial base. In the 1980s and 1990s, higher education became a priority (figure 3.1 and table 3.1). Basic education continues to be largely funded by the state, while higher education is funded primarily by individuals.

The Asian financial crisis of 1997 put the brakes on Korea's exceptional growth. Opaque financial relations between banks and large businesses covered by the government had resulted in a large volume of dubious loans. Once the situation was discovered, foreign investors fled the country, precipitating an economic crisis and a huge rise in unemployment.

To resolve the crisis and put the economy back on solid footing, the government enacted remedial financial and economic measures, while at the same time launching a nationwide, multisector KE plan (box 3.5) promoted through an awareness campaign in the nation's main business newspaper. Coordinated by the Ministry of Finance, the plan included reforms across all levels of the education system, incentives to stimulate R&D (to compensate for the business sector slowdown), the promotion of venture businesses, and the building of a dynamic information society. This last phase, the most successful of the plan, resulted in the creation of an advanced information infrastructure (as measured by Internet access, e-applications, and so on) supported by a very dynamic information technology (IT) industry.

Figure 3.1 *Enrollment Rates in Korea, 1970–2004*

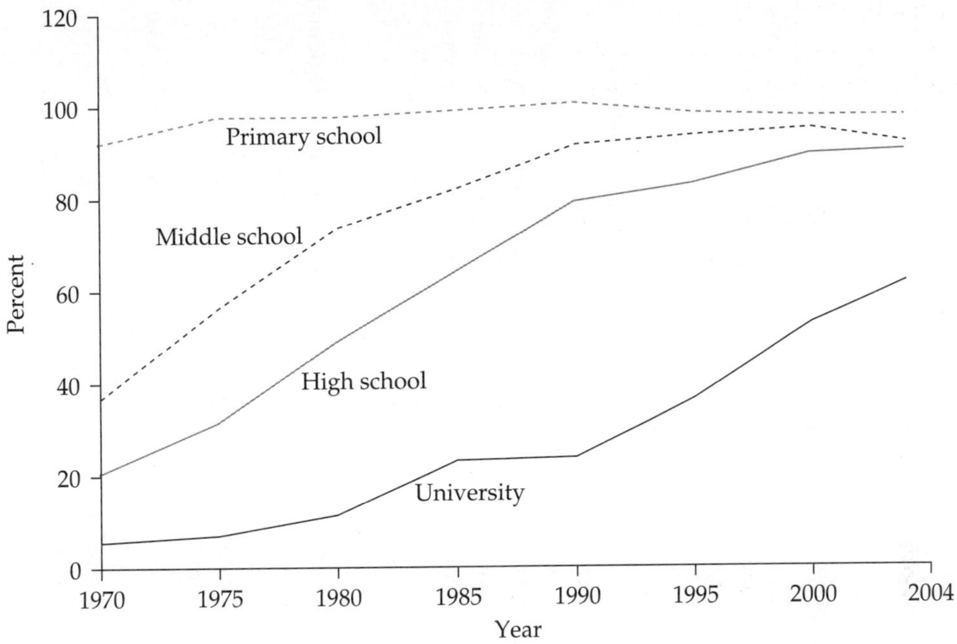

Source: Authors' calculations from World Bank internal data.

Today, Korea has maintained or recovered its top economic ranking in a number of sectors. However, it needs to encourage its technological creativity. Measured in terms of patents filed at the United States Patent and Trademark Office (USPTO), Korea falls short of its direct competitors, Japan and Taiwan (China). Korea must also address the increasing polarization between its large and small businesses, advancing and lagging regions, and the rich and poor. This requires further action to address the country's overall innovation capabilities and the effectiveness of its education, entrepreneurship, finance, science, infrastructure, and development.

Ireland

Ireland, a very poor country up to the 1950s, based its economic development on the attraction of foreign direct investment (FDI) by offering large tax incentives and subsidies. Meant to halt emigration by providing a significant number of jobs quickly, the policy was managed by the powerful Irish Industrial Development Authority (IIDA), whose budget has represented up to 15 percent of GDP.

At great expense, in the 1950s Ireland created 12 regional technology colleges throughout the country to train students to work for incoming foreign businesses. In addition, the Institute for Research and Standards was established to ensure quality, standardization, and adaptive research. A number of local offices supported the technical needs of SMEs, notably those supplying materials and components to foreign firms.

The first indigenous effort toward innovation began in the western part of the country. Shannon airport, Europe's westernmost airport for passengers crossing

Table 3.1 *Stages of Economic Development in Korea*

	Development goals	*Major policy directions*	*Macroeconomic governmental policy framework*	*Human resource development*	*Science and technology*
1960s	• Build production base for export-oriented industrialization	• Expand export-oriented light industries • Mobilize domestic and foreign capital	• Prepare legal and institutional bases to support industrialization	• Increase literacy • Establish national infrastructure	• Build scientific institutions, legal and administrative framework
1970s	• Build self-reliant growth base	• Promote heavy machinery and chemicals industries • Build social overhead capital	• Maximize growth, expand policy loans • Intervene in markets	• Increase vocational training • Increase number of engineering graduates from colleges	• Establish scientific infrastructure setting: specialized science and technology (S&T) institutions, Daeduck science town
1980s	• Expand technology-intensive industries	• Industrial rationalization • Decrease export subsidies and expand import liberalization	• Stabilization • Enhance private autonomy and competition	• Expand higher education system • Develop semiskilled labor capacity	• Promote R&D and private research center promotion • Establish national R&D programs
1990s	• Promote high-technology innovation	• Support technology development • Build information infrastructure	• Liberalization • Reform and restructure	• Develop highly skilled labor in strategic fields such as IT • Develop lifelong learning system	• Leading role in strategic area: HAN programs
2000s	• Transition to knowledge economy	• Promote venture business and small and medium enterprises	• Globalization • Balanced national development	• Increase research productivity • Improve quality of university education	• Build national and regional innovation systems

Source: WBI and KDI 2007.

the Atlantic Ocean, had undergone a serious decline in traffic owing to the development of aircraft able to fly directly to North America from other European cities. To reverse the decline, the Shannon Development Corporation was established to promote tourism, handicrafts, and cultural activities. The new agency operated autonomously from Dublin. The area also benefited from the tax advantages of Shannon airport's free zone.

To sustain growth and to respond to the population's desire for education and training, the government chartered three technological universities—or university colleges—in the early 1970s: Dublin in the central part of the country, Cork in the south, and Limerick in the west. These colleges complemented traditional universities (such as Trinity College) by offering advanced technology and management programs; they also expanded the country's academic research base. Ireland's science and technology (S&T) and IT infrastructures were further strengthened with the help of funds from the European Union.

Despite such efforts and the resulting increase in FDI, the Irish economy suffered high unemployment (15 percent of the labor force) and uneven growth through the late 1980s. This was in large part because of the overdevelopment of the government sector, which had imposed high taxes and amassed a large debt. In 1987 Ireland adopted a strong free market orientation and lessened the government's economic authority in favor of a "social partnership" agreement among the government, the business sector, and trade unions. This action gave a needed boost to the economy.

Meanwhile, the productive sector of the economy became more sophisticated. The requirements of multinational companies (MNCs) prompted Irish suppliers to improve their technologies. An expansive government program promoted the transfer of technology and management skills (see chapter 5). As foreign businesses employed a growing number of Irish graduates and established closer links with Irish university research teams, Ireland established itself as a major source of innovative solutions for MNCs. Spin-offs of such companies—created by Irish entrepreneurs—appeared. Venture capital increased. Information and communication technology (ICT) industries, largely linked to FDI, helped drive the growth process.

In recent years Ireland has strengthened its own R&D base by establishing Science Foundation Ireland with large funding capabilities. The foundation supports domestic research teams selected through an international process and helps to develop and diversify the technological and economic base of the country, notably in the fields of health and biotechnology, with new clusters in western Ireland (IIDA 2006).

Finland

Finland won its independence from Russia in 1917. Its difficult history fostered the development of a cohesive society and efficient economy that made good use of its principal natural resource: trees. Up until World War II, the Finnish economy was undeveloped, and the living conditions of its people were poor, but the country was strongly democratic and the government essentially free of corruption, valuing empowerment and education for all. During World War II, Finland was able to defend its freedom and resist invasion by the Soviet Union, albeit with some loss of territory.

After World War II, Finland took advantage of its political position to trade with the Soviet Union (largely on a barter basis) while it developed a wood-based industry (furniture factories, paper mills, and so on) and other industries of comparative advantage (such as the manufacture of ice breakers). The development of its manufacturing base owed much to the intensive use of modern technologies such as electronics and automation, and of new materials, as well as to considerable efforts in design and marketing (such as in furniture exports). By the late 1970s, Finland spent 1.5 percent of its GDP on R&D.

Regional technology-development initiatives, based on a step-by-step approach, were also encouraged by the government. With the support of the Finnish regional development agency (KERA), experimental projects in selected areas (such as the central-eastern province of Mikkeli and the northern city of Oulu) were launched. These projects included the development of technology parks, entrepreneurship training programs, and small venture capital funds. Provincial authorities were gradually granted greater budgetary autonomy for their local initiatives, creating a climate of confidence that promoted new fields outside of forest-related activities.

With the disintegration of the Soviet Union—on which Finland had depended for 30 percent of its exports—the Finnish economy suffered a severe recession. Its growth rate reversed to negative, while unemployment rose to more than 15 percent of the labor force. With the support of all political parties, the government worked to adapt the economy to the new situation. It significantly increased support for R&D despite the apparent cost to the economy—and, in the long term, the measure proved its worth. The government also systematically organized the management of a national innovation system (see chapter 5) and invested heavily in retraining programs and education. Today Finland boasts the best education performance in the Organisation for Economic Co-operation and Development (OECD).

What Finland did, in fact, was to institute a new form of industrial policy, with a shift in focus and content away from the macroeconomic approach of industrial subsidization toward the microeconomic approach of providing the right conditions (see the next section of this chapter). A very active innovation policy brought forth a series of complementary and overlapping measures, along with the establishment of a powerful national technology agency (see figure 5.1).

At the start of the last decade, enterprises in difficulty were reorganized and restructured. One case involved a company located in the central part of the country along the Nokia River. Among other things, it produced cables, television sets, and paper products under relatively uncoordinated management. In a very bold move, it decided to focus all its activities on telephone products, more particularly on mobile telephones, by taking advantage of advanced radio technologies and standards developed in the Nordic context. A robust R&D program drew on universities, public laboratories, and foreign competencies. This was accompanied by a sustained worldwide marketing and distribution campaign. Within a few years, Nokia became the world leader in mobile phones, and, in 2003, the company accounted for 20 percent of Finland's total exports and 3.7 percent of its GDP.

Various bodies—among them the S&T Policy Council, the Committee for the Future, and the Finnish National Fund for Research and Development (SITRA)—coordinated policy both horizontally across departments and vertically, from the top government level down to local constituencies, through parliament and with representatives of business and trade unions. These agencies coordinated develop-

Figure 3.2 *The Stages of Industrial and Economic Development in Finland*

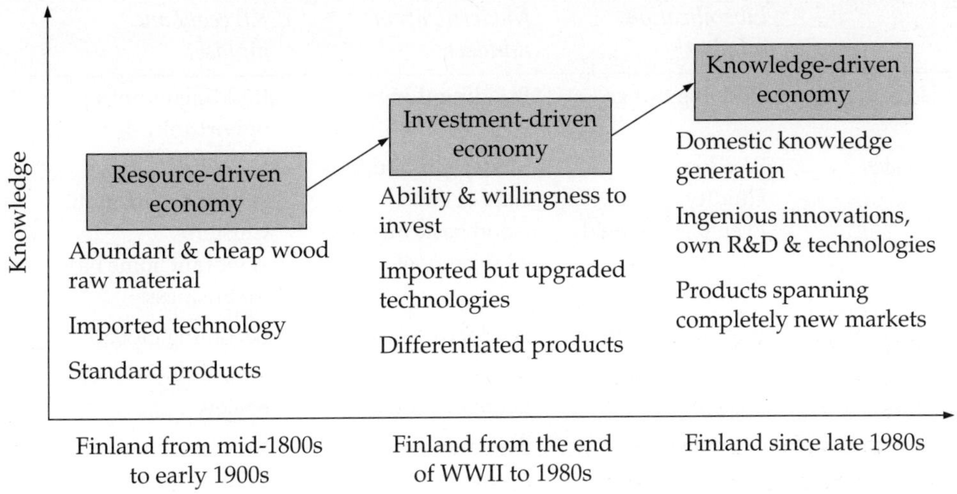

Source: Adapted from Dahlman, Routti, and Yla-Anttila 2006.

ment strategies and the evaluation of policy models with the help of foreign experts (box 3.4).

Figure 3.2 summarizes industrial and economic development in Finland since the mid-1800s.

Inspired by these examples, the following sections set out principles for implementing KE strategies, including:

- the change of mindset needed for KE strategies
- the general attitudes that should inspire KE strategies
- the adaptation of policy measures to development levels
- the management of reform
- the exploitation of entry points such as driving sectors and cities
- the need to deal with contextual specifics of various types.

Adopting Conducive Attitudes

A New Mindset for Government Action

As shown by the examples of the Republic of Korea, Ireland, and Finland, KE development calls for government action beyond the familiar programs of market liberalization and selective, modernizing interventions. Table 3.2 summarizes this new approach, in which the government must challenge established interests and institutions, stimulate change and initiatives, and act as a policy integrator. The new approach complements, rather than replaces, the liberalization and modernization views.

Key Attitudes

The examples of Korea, Ireland, and Finland also illustrate the general attitudes that should guide knowledge-based strategies: determination, vision, openness,

Table 3.2 A New Mindset for Government Action

	Liberalization mindset	*Modernization mindset*	*KB economy mindset*
Is about	Undoing things	Building things	Building winning opportunities
Creates	Freedom Fluidity Even playing field	Modern institutions Rule of law Good basic business environment	Vision A winning mentality Clusters A vibrant home base for business
Main Focus	Stability, incentives	Productivity catch-up	Becoming globally competitive
Domain	Economy	Economy, society	Society
Government's role	Get out of the way Stop being an operator	Become a good regulator	Become a challenger Become an integrator

Source: Adapted from Rischard 2002.

and pragmatism. The same attitudes underpin the successful efforts of other countries as well: Vietnam's entry into the global knowledge economy (box 3.1); Estonia's drive to make each citizen IT-literate by encoding this objective in its national constitution (see chapter 6); and Tunisia and Malaysia's bold national KE plans that catalyzed their development (see chapter 8 on active reformers).

DETERMINATION. A KE-based approach requires determination. Adherence to the so-called Washington Consensus on policy reform—which calls for macroeconomic stability, deregulation, trade liberalization, and privatization—is not sufficient in itself. Policies need to address all intangible assets and sources of growth—education, research, information, communication, and entrepreneurship—in order to foster and apply knowledge throughout the economy.

Determination requires thinking big. Successful knowledge-based strategies require determined action across sectors and fields. Restricting efforts to a specific policy plank is thinking small.

Determination involves the ways and means used to accomplish the basic policy actions needed at a nation's stage of development (as discussed in the next section). While it is difficult to make the transition to a stage of higher development, it is possible to apply modern means to achieve the objectives applicable within a stage. For instance, the use of advanced communication tools and distance learning can facilitate the meeting of education objectives even in the poorest countries. Similarly, the use of basic telephone and Internet facilities in countries at a higher stage of development can rapidly and radically transform the conditions within which entrepreneurs—including farmers and fishermen—do business. The application of these means can be effectively supported by government efforts.

Determination is demonstrated by clearly structured industrial policies set to facilitate the development of a strong manufacturing sector. These measures improve the overall environment in which businesses evolve.

VISION. Countries that advance have started with a vision that, in one way or another, points to a goal and gives a sense of identity. A clear vision gives expression to determination. A vision generally takes the long view, sometimes with the fruition of goals 20 years out.

Vision arises from small groups of people, from community or regional leaders, and sometimes even from the head of state. Visionaries need to look for resources in various sectors of society, such as business and education. This is necessary to anchor the vision in reality and to obtain the commitment of the populace. It is crucial to realize a vision in concrete terms quickly—in tangible projects, even if of modest size. The vision thus becomes credible and reinforces national investment and self-confidence.

OPENNESS. Another lesson from the Korean, Irish, and Finnish experiences—and from other successful transitions to a KE approach—is the need for openness to the outside world. Globalization offers considerable opportunities; chief among them is the opportunity to attract FDI and employ it appropriately. Each country must organize instruments and channels to systematically monitor technologies and knowledge abroad that might be relevant to its activities and goals.

Countries that have experienced significant emigration can use their diaspora communities as a resource. Emigrants abroad can help establish close technology and business links with their home country (see chapter 7). India is an excellent example of the benefits to be gained from such links. As illustrated by the experiences of Scotland and Central European countries, emigrants can also be a source of expertise for policy making and business development. Countries may also offer incentives to lure highly educated people back to their home country, where they may help to develop new industries, as in the cases of Taiwan (China), Ireland, and Israel.

A successful knowledge economy relies on policy exposure, which can be gained through international exchange, study tours, and pilot programs based on policy measures that have proven successful abroad.

PRAGMATISM. Determination, vision, and openness must be grounded in reality. Policy makers need to clearly understand the needs and constraints of their economy and temper their ambitions and goals, adapting their efforts to their country's capacities and resources. They must make the best use of their country's competitive advantage, whether in agriculture, tourism, or natural resources, and to direct their attention first to the areas with the highest leverage to position the country on a successful KE track. As the experiences of the Republic of Korea, Finland, and Ireland demonstrate, building a knowledge economy is a gradual process in which efforts, investments, and policy actions are adapted at each stage of development, accompanied by an understanding of the country's specific needs, capabilities, and comparative advantages.

Adapting Policy Actions to Development Levels

The three examples of the Republic of Korea, Ireland, and Finland suggest policy actions appropriate for various stages of development. Progress toward a knowledge economy is measured in relation to the stages of development as defined by the World Bank, and by respective levels of advancement. Low-income countries are at an *early* KE stage and need to build foundations; lower-middle-income countries are at an *upgrading* KE stage and need to raise their KE assets before they can embark on a broad KE strategy for growth; upper-middle-income countries are at an *emerging* KE stage; and high-income countries are ready for a *full-fledged* KE strategy. See table 3.3 for an overview of KE-supporting governmental actions for each of the four stages.

Low-Income Countries

Low-income countries at an early KE stage need to establish solid foundations in governance and the business environment. Governments may choose to establish special economic zones (SEZs) with few bureaucratic entanglements and transaction costs. This attracts foreign investment, which introduces new technology and management and creates jobs. Large but vital tasks are (a) the reduction of illiteracy through basic education and (b) the strengthening of a few technical and tertiary institutions to build core competency in advanced technology, engineering, and science. For ICT advancement, low-income countries should first build a minimal telephone infrastructure that takes advantage of mobile technology and then establish fixed-line connections for the Internet (at least 10 percent of the population must be connected in order for the knowledge economy to take off). For educational and cultural advancement, they should also make good use of TV and radio networks, notably to reach rural areas. In terms of innovation, they should make the best possible use of national and global knowledge to serve the basic needs of the population (for food, healthcare, and housing), and develop basic infrastructure for quality control, metrology, and other services essential for supporting technology diffusion and adaptation throughout the nation, particularly in rural areas. Investments may be directed to selected IT niches if it is possible to take advantage of a literate labor force and entrepreneurial individuals well connected to international markets. The case of Vietnam illustrates how a low-income country's policy may embrace a KE approach (box 3.1).

Lower-Middle-Income Countries

Lower-middle-income countries that are upgrading toward a knowledge economy should further improve their business environment by focusing on financial and labor markets and by facilitating the reallocation of both financial and human resources toward an emerging formal private sector. Bureaucratic and regulatory obstacles that prevent expansion should be removed. SEZs should be developed across the economy, and more FDI attracted through targeted strategies and incentives. To achieve full literacy and expand the higher education base by joining networks of advanced institutions worldwide, there must be full access to primary education and increased standards of quality as well as access to secondary and vocational education. Internet access should be expanded to improve governance, logistics, business services, and the delivery of social services. Innovation requires an increased awareness of global developments to identify and import relevant technologies. Extension services designed to increase productivity in agriculture and manufacturing should be increased. While private R&D may be encouraged, the existing public R&D infrastructure should be strengthened. Both must be sup-

Box 3.1 *Vietnam's Development and KE Elements*

Vietnam was able to tap into global knowledge through trade liberalization, foreign direct investment supported by institutional reforms (notably property rights in agriculture), and the encouragement of private investment. Through more efficient use of public funds in education, primary enrollment rates increased to nearly 100 percent of the eligible population, and secondary enrollment rates expanded from 32 percent in the early 1990s to 70 percent in 2002. Thanks to a strong market-focused reform agenda, Vietnam's average annual GDP grew 5.3 percent from 1993 to 2002.

In the 1990s, the government decentralized economic management; explicit subsidies to state-owned enterprises (SOEs) were reduced, and the number of SOEs decreased from more than 12,000 in 1990 to around 6,200 by 2002. Productivity levels rose by an average of 11 percent a year from 1997 to 2002. The government replaced various administrative measures with economic measures, including a market-oriented monetary policy. It allowed state- and privately owned industries to deal directly with foreign markets. Vietnam's export policy led to a rise in trade from around 60 percent of GDP in the early 1990s to more than 100 percent in 2002. Following the 2001 bilateral trade agreement with the United States, trade between the two countries doubled in both 2002 and 2003.

One-third of the total population of Vietnam has emerged from poverty since the early 1990s: the percentage in poverty fell from 58 percent in 1993 to 29 percent in 2002. This occurred initially because of the redistribution of agricultural land to rural households, and later because of the modernization of agriculture and the development of private enterprise. Industry is now the driving force behind growth in the areas of food, mining, chemicals, textiles, and clothing. Industrial growth has been funded largely by foreign investment in joint ventures and foreign-owned subsidiaries. Industrial output rose by nearly 17 percent annually between 1999 and 2003. In traditional production, Vietnam is the second-largest Asian exporter of rice and coffee.

The hardware and software industries have grown steadily thanks to rising demand abroad. Japan is a major importer of Vietnam's software products. Hardware production grew by 27 percent between 2002 and 2003, and software production by 40 percent. By the end of 2003, there were 2,500 registered ICT businesses—including 570 software companies—and 12,000 ICT workers. The ICT sector contributed 0.3 percent to GDP growth from 1993 to 2001. The government wants to build up clusters and software parks in Ho Chi Minh City and Hanoi. Ho Chi Minh City boasts online education facilities, a ready ICT workforce, local software development operations, companies that manufacture personal computers, and a university and colleges.

Sources: World Bank 2005b, Chidamber 2003.

ported by measures to increase technological and managerial competence. University-industry interaction should be encouraged on a selective basis through appropriate support and incentives.

The example of Botswana illustrates the steps that a lower-middle-income country can take in upgrading toward a knowledge economy (box 3.2).

Upper-Middle-Income Countries

As they move closer to a solid knowledge economy, upper-middle-income countries should further strengthen their business environment. In particular, they must focus on financial and equity markets by facilitating the mobilization of development and

Table 3.3 *KE-Related Government Action at Four Development Stages*

	Economic and institutional	*Education and training*	*ICT*	*Innovation system*
Low-income	Strengthen basic institutions of governance, administration; expand property, finance. Liberalize access to global goods and services. Reduce barriers to trade within the economy. Encourage FDI. Establish SEZs that reduce bureaucratic problems and transaction costs.	Enhance quality of and expand access to basic education to reduce illiteracy. Strengthen key secondary (technical) and higher education institutions to build a core nucleus of advanced competences.	Extend radio, television, and telephone service (with fixed-line service for Internet connection) with focus on expanding access to rural areas.	Strengthen capacity to scan, acquire, and adapt global knowledge to local needs. Blend global with local knowledge for relevant application. Focus on dissemination of knowledge to upgrade traditional sectors. Develop basic infrastructure for metrology, quality control. Use knowledge to enhance value of natural resources. Establish demonstration projects of innovative technology for local settings.
Lower-middle-income	Strengthen financial and labor markets to improve business environment. Extend SEZs across the economy. Encourage more strategic FDI.	Enhance quality and expand access to secondary and vocational education. Strengthen key higher education institutions by integrating into networks of advanced institutions worldwide.	Expand the use of the Internet to improve governance, logistics, business services, and delivery of social services.	Intensify global scanning to find, acquire, and import relevant technologies. Increase productivity and agricultural extension services. Strengthen existing public R&D institutions and encourage private R&D with measures that increase technological and managerial competence.

Table 3.3 *Continued*

	Economic and institutional	*Education and training*	*ICT*	*Innovation system*
Upper-middle-income	Strengthen financial and equity markets by facilitating the mobilization and development of venture capital. Improve the efficiency of government, tax collection, and expenditure with improved governance. Expand IPR protection.	Continue to improve the quality of secondary education. Expand access to and quality of higher education. Develop lifelong learning systems enabled by multiple pathways and providers.	Further develop and diversify the application and use of Internet-based technologies to reduce transaction costs and improve the overall efficiency of the economy.	Encourage domestic innovative capacity with appropriate incentives. Encourage the private sector R&D. Aim to increase domestic R&D expenditure to 2% of GDP.
High-income	Ensure the responsiveness of government institutions. Strengthen IPR enforcement. Strengthen investments in intangibles such as R&D, education, software, marketing, distribution, and high-value knowledge services.	Strengthen higher education sector as part of a seamless lifelong learning system, where a large percentage of tertiary students are adults returning for specialized training and certificate programs.	Use ICTs strategically as key infrastructure with development of special applications software and media convergence.	Increase domestic R&D expenditure to 3% of GDP. Strengthen strategic international alliances for R&D, production, marketing, and distribution. Target support toward specific clusters.

Source: Authors.

Box 3.2 Botswana: Investments Supporting KE Development

Botswana emerged as a middle-income country with a per capita GDP of $3,684 in 2004. Its economic achievements are due to significant investment across all four pillars of the KE approach. These investments were facilitated by Botswana's position as the world's largest exporter of diamonds; nonetheless, the government's long-term development planning was crucial in designating funds for investments that promoted growth and human development while maintaining fiscal discipline.

The achievements that are most apparent are related to the economic and institutional regime. Policies governing trade, banking, exchange rates, and fiscal practice were all reformed. As a result, Botswana was recently ranked 30th worldwide in economic freedom and received an "A" grade from Transparency International as the country with the best credit risk and least corruption in Africa (see chapter 4 for more detail on the EIR reforms undertaken).

Perhaps less apparent, but a definite boon to longer-term development, have been the investments made in education. Adult literacy rates have increased dramatically since 1999. The primary gross enrollment ratio has stabilized at around 100 percent, the secondary ratio has improved from 40 percent in 1990 to 70 percent, and tertiary enrollment rates have steadily increased, although the latter remain significantly lower than those of the regional leader, South Africa.

Finally, ICT and related investments have increased significantly, particularly in fixed-line and mobile phones and personal computer access. Although the number of Internet hosts remains much lower than in South Africa, Botswana has invested far above the African average.

The results have been impressive: real incomes have risen across the economy, and although relative income inequality has not been reduced, it has not worsened. With real per capita GDP growth rates averaging nearly 8 percent over the past three decades, the poor are better off than before. Unemployment rates are low by Sub-Saharan African standards, although youth unemployment has recently emerged as a serious problem, and the country continues to suffer a shortage of skilled labor. Its record in human development is also impressive, with one vital exception—HIV infection. The rapid spread of HIV/AIDS may have substantial implications for developing human capital and improving productivity.

Sources: World Bank 2000a; WEF 2003, U.S. Department of State 2005, IMF 2004, Land 2002, Transparency International 2005.

venture capital. The efficiency of government tax collection and expenditure should improve with an educated labor force and improved governance. Access to higher education should continue to widen and the quality of education to improve. Lifelong learning systems characterized by multiple pathways and providers should be developed. The application and use of Internet-based technology should be further developed, increased, and diversified to further reduce transaction costs and improve economic efficiency. Domestic innovative capacity should be encouraged through appropriate incentives (reimbursable subsidies, tax incentives, and so on), particularly for developing private sector R&D, with a goal of increasing R&D expenditure to 2 percent of GDP. Protection for intellectual property rights (IPR) should also be expanded, although this is less important for low-income countries, some of which may even be penalized by the stringent IPR systems mandated by the agreement on Trade-Related Aspects of Intellectual Property Rights (TRIPS) adopted at the end of the Uruguay Round of world trade talks in 1994 (see chapter 5). The experience of Chile illustrates appropriate policies for upper-middle-income countries (box 3.3).

> ***Box 3.3*** *Chile: Exemplary KE Efforts, but a New Push Needed*
>
> Chile boasts an excellent economic and institutional regime. It is ranked 13th in the Economic Freedom Index and 20th—one place after the United States—in Transparency International's Corruption Perception Index 2005. Chile has also made the necessary investments to strengthen functional knowledge pillars.
>
> Through the successful and timely employment and coordination of economic, institutional, and technological resources, Chile developed a highly dynamic export sector market for its wine, salmon, and paper industries. Institutions such as Corfo and Fundación Chile (see chapter 5) played a critical role in this development, particularly for the salmon industry.
>
> However, Chile's formal R&D base and industry-research collaboration remains relatively weak. The nation seems trapped in a natural-resource-based economy. How can it free itself? First, Chile must strengthen its science and innovation system. Secondly, it must increase the general education level of the population (see chapter 8 and Benavente, de Mello, and Mulder 2005).

Advanced Countries

For advanced economies, development and maintenance of a true knowledge economy require an immediately responsive and flexible environment. Incentives should be directed toward intangibles such as R&D, education, software, and marketing and should be adapted for a service-based economy. In the education sector, the priority should be to increase access to and quality of the higher education sector. This, in return, becomes part of a larger, seamless, lifelong learning system with a large number of tertiary students, including adults. ICT becomes the basic infrastructure of the economy with a broad development of special applications, including dedicated software and multimedia. Innovation becomes the key engine of growth; the domestic R&D effort should reach 3 percent of GDP. International strategic alliances for R&D, production, and marketing are encouraged by government support.

Managing Reform Processes

Timeline and Impact of Reforms

KE reforms can have a very significant impact in a relatively short time, even though their full effect requires sustained action across the four pillars. The effect of measures that improve the business environment may be felt in one or two years—sometimes in only a few months—in areas such as enterprise development and the attraction of FDI. Similarly, investments or actions relating to ICT may show tangible effects in only a few years—witness the rapid spread of cell phones. By contrast, innovation policy requires a minimum of five years to generate significant improvements in technology diffusion, job creation, enterprise growth, and international competitiveness. And education policy reforms will not take full effect until the passing of one to two decades at best. However, measures to retrain workers—and more generally to establish lifelong learning venues—should improve employment opportunity for much of the population far more rapidly.

KE development processes are nonlinear. Unexpected events—such as a crisis that demands immediate decisions or the restructuring of a sector or firm that leads to immediate and unanticipated industrial growth—can effect a major change in direction. As mentioned above, following the collapse of the Soviet Union, a major buyer of Finnish goods, Nokia focused all of its production on the manufacture and marketing of mobile phones. In a decade, the industry came to account for some 20 percent of the country's total exports and 3.5 percent of its GDP. At the same time, the national innovation policy was reshuffled and strengthened—an illustration of the power of voluntarist policies in successful KE stories.

Knowledge Dynamics: Incremental Change

As previously discussed, determination and vision are necessary to build confidence that a new and better era in national development is at hand. However, the conditions for substantial change throughout the institutional system are often not fulfilled even in countries that have been affected by a deep crisis. When effective market mechanisms and government organizations are in their infancy, policy makers may face both market failure and government failure. Under such conditions, pragmatism—adopting and adapting what works—should inform knowledge strategies.

The design of institutional solutions for knowledge-based growth does not require full-scale public sector reform. If resources are few and time is constrained, policies that establish institutional shortcuts may be appropriate. Imperfect and idiosyncratic institutions may ensure a functional fit between a country's conditions and the challenges of reform.

For example, many observers have been puzzled by the remarkable success of town and village enterprises in China. These enterprises were owned and controlled by local governments. Standard theory cannot account for their comparative advantage over private enterprises. It seems that the public structure accommodates the particular features of the Chinese economy and society at this point in time. China is not the only country to employ incremental reform. Modest reforms appeared to account for economic growth in India, allowing the nation to exceed its traditional growth rate of 3 percent. In the 1980s, under Rajiv Gandhi, the government relaxed industrial regulations, encouraged imports of capital goods, and rationalized the tax system. Though the reforms were modest, they tipped the balance by encouraging rather than discouraging entrepreneurial pursuits. Entrepreneurship is both a principal route into global knowledge flows and a principal actor in transforming knowledge into wealth.

The recent surge of growth in these emerging giants (see chapter 8) can be traced to their strategy of gaining knowledge that can then be transformed into wealth. The reforms in China and India illustrate incremental changes from the bottom up, offering a favorable balance of risks and returns by encouraging first steps at many and diverse entry points. This incremental process increases the chances of setting the cycle of institutional reform and knowledge-based development into motion.

Sustaining Knowledge Dynamics: Bottom-Up and Top-Down Initiatives

As previous sections have shown, most developing countries need to implement major reforms if they are to move ahead. Developing a consensus for reform agendas can be as challenging as removing the institutional impediments to reform. In a pioneering study, Olson (1982) showed how distributional coalitions block, change, and stifle growth, and how new crises destroy old coalitions.

Box 3.4 Key Bodies for KE Reforms in Finland

The Science and Technology Policy Council, established in 1990, is chaired by the prime minister and composed of key ministers, including the minister of finance. The council coordinates innovation policy for all concerned departments. It was the world's first body explicitly to embrace the notion of a national innovation system to define and analyze policy.

The Finnish Parliament's Committee for the Future initiates broad public discussions about upcoming trends and future challenges, particularly those that are technology related. In the early 1990s, the committee undertook a forward-looking study involving consultations with foreign institutions worldwide. Its mission is to be aware of—and to make society as a whole aware of—major issues confronting the country.

SITRA is a parliamentary agency established in the 1960s to promote innovative initiatives for the benefit of the economy and of society. In the 1980s, it initiated Finland's first venture capital funds. It has also developed high-level training sessions in economic policy for members of parliament and key decision makers (such as top business managers and trade union chiefs), thereby developing an understanding across the economy of changes needed and of requirements for sound economic development.

Source: Authors.

Finland and the Republic of Korea are good examples of concerted consensus-building efforts to engineer successful transitions to knowledge-based economies. In both cases, a national economic crisis compelled the affected actors to define and implement a new agenda through explicit or implicit national consensus on goals and mechanisms for moving forward. Policy makers and private sector leaders extended the time horizon for results from the adopted policies. In both cases, mechanisms already in place anticipated change and the need to undertake or adjust appropriate reforms. These cases show that to overcome institutional rigidities and bottlenecks, a combination of top-down and bottom-up policies is necessary.

In the Finnish approach, agencies such as SITRA and the Committee for the Future prepare parliament and the populace for change and encourage patience with a long-term vision. Other institutions such as the Science and Technology Policy Council facilitate effective interdepartmental coordination. These bodies encourage a sustainable, long-term commitment to reform while coordinating top-down and bottom-up initiatives.

In Korea, initiatives were coordinated by the government after being set in motion by private sector initiatives (box 3.5).

Sequencing Reforms

Transitions are required to facilitate the concerted efforts that are crucial to successful reforms. Inspired by successful processes, one may propose a three-stage scheme:

- *Immediate agenda.* Through a top-down initiative, create awareness, develop rational indicators to monitor progress toward a knowledge economy, and evaluate ongoing pilot initiatives.
- *Short- and medium-term agendas.* Through top-down and bottom-up cooperation, institute a shared vision led by the private sector, institute a national

Box 3.5 *Korea's KE Transition: Bottom-Up Initiatives and Government Action*

In 1998, in the wake of the Asian financial crisis, the Republic of Korea officially launched a national strategy to move toward a knowledge economy. The initial impetus came from the *Maeil Business* newspaper in the private sector. The newspaper argued in 1996, even before the crisis, that a more coherent vision of the future of the Korean economy was urgent, and it promoted its "Vision Korea Project" in a national campaign in February 1997 and developed the first *Vision Korea Report*.

The government's ministry of finance and economy eventually championed the KE policy agenda. The Korean Development Institute, a so-called system integrator, coordinated the work of a dozen think tanks. A joint World Bank and OECD report (Dahlman and Andersson 2000) provided a framework by outlining concrete steps for reforms in various policy domains.

Progress was monitored closely to identify and address inertia or resistance. Korea's knowledge strategy of April 2000 evolved into a three-year action plan focusing on five main challenges: developing information infrastructure, expanding human resources, developing knowledge-based industry, enhancing science and technology, and eliminating the digital divide. To implement the action plan, Korea established five working groups involving 19 ministries and 17 research institutes. The ministry of finance and economy acted as coordinator. Every quarter, each minister submitted a self-monitoring report to the ministry of finance and economy, which in turn published an integrated report detailing progress. Mid-term results and adjustments to the three-year action plan were regularly implemented.

Source: Authors.

Figure 3.3 *Time Sequencing to Create a Virtuous Cycle of Change*

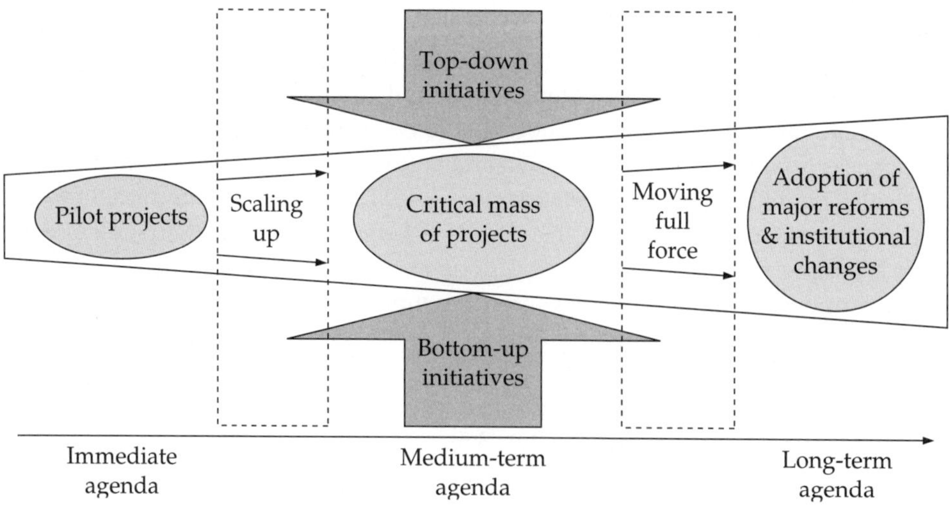

Source: Authors.

monitoring system linked to budgetary priorities, and consolidate micro-level "rapid results" projects and/or pilot projects in visible initiatives across regions and sectors. The priorities of a national monitoring system can be expected to result in significant changes in budgetary priorities.

- *Longer-term agenda.* Set a full-fledged reform agenda that will eliminate or transform major vested interests and will introduce a new incentive structure for major agents.

Exploiting Entry Points: Driving Sectors and Cities

Innovation and growth often arise in specific sectors or locations following the accumulation of a critical mass of talent, resources, and entrepreneurship. There must be an adequate and functioning infrastructure (power, transportation) in place, and a permissive—if not supportive—environment for entrepreneurial initiatives. When these conditions coalesce, competitive industries emerge or clusters develop. There are many examples of this process in advanced countries; the Irish Shannon-Limerick area and Finnish cities are cases in point. There are many examples to be found in lower-income countries as well.

The role of government is to facilitate innovation and growth by bringing together the elements and personnel that can make a difference. In its pragmatic approach (discussed above), China intentionally created enclaves for growth known as export processing zones (EPZs) and technology parks within SEZs that offer financial and regulatory incentives to local and foreign enterprises willing to relocate, along with training facilities (see chapter 4). Starting with low-skill manufacturing and progressing to astute competitive schemes, EPZs have been established along the Chinese coast.

Well-equipped government laboratories or state schools led by visionary leaders and accompanied by an active private sector provide an efficient nucleus for clustering processes. Two successful examples from Latin America are Brazil's Campinas and Mexico's emblematic Monterrey. However, Latin American governments have been unable to stimulate the replication of these successes and thus to create a critical mass of successful educational institutes that would have a significant macroeconomic impact on the region.

Sometimes innovation and growth emerge with little government participation. The city of Bangalore in India offers an example. It began as an active IT service center, drawing on local IT schools and a few private enterprises that had contracts with U.S. firms located in Silicon Valley. With a well-trained and cheap labor force, it grew rapidly. Bangalore now seems to be reaching certain limits, but its success has been emulated by other Indian cities.

More generally, IT communities and sectors are plausible entry points throughout the world. They are led by entrepreneurs using new technologies and offering attractive opportunities for employment, for profit, and for exports within a relatively short time. In today's world, ICTs appeal to the public at large and offer an opening into the knowledge and information age. As a result, governments are happy to entertain ambitious plans proposed by IT lobbyists and to devote significant resources to support their causes.

Dynamic entry points can affect an entire region, as in the case of Dubai, which founded its development over two decades on a clear knowledge and innovation strategy. With a vision articulated at the highest level of government, sustained by new and renewed investments, and operated by strong agencies, Dubai has developed an exciting economy. It has been able to diversify gradually from its shrinking base of oil reserves by first developing a world-class port and then attracting tour-

Box 3.6 *Dubai's Transition to a Knowledge Economy*

The economy of Dubai and the other six emirates that compose the United Arab Emirates (UAE) was long based on pearls, fishing, trade, and, to a limited extent, agriculture. The discovery of oil off Abu Dhabi in 1958 marked a turning point. However, although Dubai controlled the second-largest reserves after Abu Dhabi, its resources were significantly less. By 1991, Dubai's leaders realized that production and revenue would begin to decline markedly after 2010.

For Dubai to continue to prosper, diversification was necessary. Two successive generations of visionary leaders implemented a multiphase, multidimensional program to advance Dubai to the forefront of the knowledge economy. The aim of the first phase was to transform the emirate into a transportation and logistics hub linking South Asia, the Middle East, and Africa. This led to the creation of the world's largest dredged port (currently ranked 10th in terms of container ship passage), a world-class airline (Emirates), and, after the success of the Dubai International Airport, the construction of a second airport—the world's largest—Jebel Ali Airport, which sees 120 million passengers (compared to Heathrow's current 85 million) each year. The transport hub strategy led to a tourism strategy that, against all expectations, has been a noteworthy success. Over 15 million tourists visited Dubai last year, and the emirate has built several high-profile resorts and attractions (including Dubailand—the world's largest shopping mall) to keep the tourists coming.

The aim of the second phase of Dubai's development was to develop core competencies in technology, media, telecommunications, e-commerce, and other specialized domains. This strategy, launched by Dubai's current leader, Sheikh Mohammed bin Rashid al-Maktoum, centered on two successive plans: the 2010 and 2020 Vision Plans. These plans provide well-designed financial and economic incentives for companies in a media free zone (TECOM) that consists of three separate entities: Dubai Internet City, Dubai Media City, and Knowledge Village. The TECOM complex boasts the Middle East's most extensive IT infrastructure and the world's largest commercial Internet protocol telephony system. Dubai Media City opened for business in January 2001 and has since attracted some 850 companies. Knowledge Village provides world-class training services in management, business, human resources, and education support. Other specialized technology zones were or will be created. These include Dubai Technology Park, Dubai Biotechnology and Research Park (DuBiotech), and Dubai Industrial City (DINC).

The aim of the third and latest phase of Dubai's development is to establish the emirate as an international financial hub. This has been facilitated by the massive return of Middle Eastern finance closer to home following the attacks of September 11, 2001, and by the recent boom in oil prices. After some wrangling with the UAE's central authorities, Dubai won permission to exempt its financial center from almost all of the federation's commercial laws and to establish a separate, Western-based commercial system that would do business in dollars and in English. Independent Western regulators and judges were invited to the emirate, a move that is particularly important for success as an international financial center. However, this third phase has not been without difficulties, which nearly led to the collapse of the entire project in 2002. When property development rights in the financial center were allocated to local businessmen outside of the designated auction process, even though there was no suggestion of foul play, the Western regulator made it plain that anything even *perceived* as backroom dealing would undermine the center's reputation. The Dubai financial center reacted by sacking the regulator. This nearly caused a stampede among the international financial firms present. However, the project was put back on track.

Source: Authors.

ists. It established Internet and media connectivity in cities and is now attempting—not without some difficulty—to establish itself as a financial hub. Its economy has doubled in real terms in the last 10 years. The declining importance of oil production in the economy is illustrated by the fact that oil accounted for 20 percent of GDP as recently as 2001, but now accounts for less than 10 percent. Less than one-fifth of the local population are from the United Arab Emirates (UAE), which demonstrates that Dubai has succeeded in attracting global players to its various cutting-edge domains. Dubai is contributing to a knowledge and technology renaissance throughout the Middle East (box 3.6).

The successful creation and scaling up of entry points encourages reform by overcoming resistance to change, convincing skeptics, and increasing confidence. This is a major reason to work for their success. When entry points do not succeed, concerned communities may come to doubt the entire initiative, particularly if huge investments in ambitious projects do not bear fruit as early or to the degree expected.

Dealing with a Country's Context

Development Trajectories and Policy Agendas

The World Bank has recognized the need to adapt development strategies and policy measures to each country's specific context. When considering the development trajectory that is most appropriate for a country, it is crucial to consider different approaches in industrial strategies, as seen in the three examples above. Korea developed its industries with technology from abroad, through its licensing policy and systematic OEM agreements. The core *chaebol* industry groups were family owned. Ireland actively solicited FDI for its development, while Finland, in contrast, developed a vibrant indigenous R&D base and did not solicit foreign investments for its industrial development. These three models represent typical strategies adopted by advanced countries. In chapter 8, the important differences in development trajectories and related policy agendas in middle- and low-income countries will be discussed.

Challenges and issues therefore differ with regard to today's policy agendas. For example, Korea should now expand its indigenous innovative capability and concurrently address the trends toward polarization of its economy and society. Ireland should build a larger research base and diversify its innovation clusters. Finland should maintain its position of technological pioneer and world competition leader by finding new niches.

In recognition of the significant differences between countries, the World Bank recently tested a growth diagnostics methodology based on the identification of binding constraints. This method, expounded by leading development economists (Haussman, Rodrick, and Velasco 2005), identifies the key bottlenecks to growth processes. A government must focus its policy actions on removing these systemic obstacles rather than employing the usual laundry list of measures that touch all areas (trade, investment, finance, governance, labor, and so on).

The authors illustrate their theory with the examples of Brazil and El Salvador. They conclude that the growth process in Brazil is affected in the first instance by constraints on entrepreneurs—particularly the lack of development capital. By contrast, in El Salvador the major constraint seems to be a lack of ideas, and of knowledge more generally. The situations thus require very different policy approaches.

(Note that both approaches adhere to the KE philosophy.) As confirmed by World Bank analysts, application of the growth diagnostics methodology requires a thorough understanding of the country concerned.

Sociocultural Issues

The remainder of this section offers insights on addressing culture-related topics. Sociocultural considerations are of paramount importance in the development process. Whatever the policy actions and strategies for change, slowly changing sociocultural specificities will shape efforts, investments, and growth trajectories.

Cultural influences on and implications for countries' economic systems and policies, particularly their knowledge and innovation dimensions, can be approached at the different levels of a "culture tree": broad civilizations, subcivilizations, and nations (Aubert and Landrieu 2004).

There are striking differences between Eastern and Western civilizations. These can be imputed in part to different cognitive processes (box 3.7), with implications for relationships to the world, as well as societal organization. Two different postures can be identified: in the West much thinking involves a distancing from reality, in the East an immersion in it (Aubert and Landrieu 2004). These different

Box 3.7 *Differences between Eastern and Western Cognitive Processes*

Is it possible that the basic human activity of thinking, which until recently was thought to be universally homogenous (all people reason and think the same way), differs, in fact, according to the cultural context?

Richard Nisbett, an eminent cultural psychologist, demonstrates in *The Geography of Thought* that Asians (defined broadly by the author as people from "China and the other countries heavily influenced by its culture, most notably Japan and Korea") and Westerners "have maintained very different systems of thought for thousands of years," and that "human cognition is not everywhere the same." He argues, in a nutshell, that those brought up in Western and East Asian cultures think differently in scientifically measurable ways.

This difference is verified in experiments and data: East Asians are measurably more holistic in their perceptions (taking in whole scenes rather than a few outstanding objects); Westerners focus much more on identifying and remembering what is prominent in certain scenes. Nisbett explains these differences as "an inevitable consequence of using different tools to understand the world."

A brief historical overview sets the backdrop for the differences between the two groups: an overview of ancient Greek society and Aristotle, and ancient Chinese society and Confucius shows the practices that produced these differing approaches to the world. Westerners tend to inculcate individualism and choice, while East Asians are oriented toward group relations and obligations. "The individualistic or independent nature of Western society," writes Nisbett, "seems consistent with the Western focus on particular objects in isolation from their context, and with Westerners' belief that they can know the rules governing objects, and therefore, can control the objects' behavior." By contrast, the "collective or interdependent nature of Asian society is consistent with Asians' broad, contextual view of the world, and their belief that events are highly complex and determined by many factors."

Source: Nisbett 2003.

ways of thinking imply differences in various domains of human activity including medicine, law, science, human rights, and international relations. In science and technology, the Western approach to reality favors a scientific search for causality in understanding natural phenomena, while the Eastern mind favors holistic combinations of existing elements as the basis for technological development. With regard to the legal and institutional environment, Western societies are concerned with the establishment and observance of the rule of law as the basic means of protecting the individual, while Eastern societies tend to emphasize informal relationships regulating collective groupings, such as the Chinese *guanxi*. This leads to two clearly different economic systems with some contrasting features (table 3.4).

The historical experience of nations, and their geographic location, also play a vital role in shaping collective mindsets and behaviors. At the level of nations, behavior and thinking are strongly influenced by history. For the developing world, the impact of colonization is particularly important. The situation is better when trauma has been limited or the contact has been well integrated. Japan, for example, has maintained its integrity throughout its history, and has thus been able to integrate modern features into its traditions. Botswana is another instance in more recent times, and on a particularly troubled continent. As far as geography is concerned, an island—in geographic and cultural terms—seems to possess a special sense of identity that helps to mobilize the available resources, provided that the country is open enough to external pressures and opportunities (Aubert and Chen, forthcoming).

All value judgments should be eliminated. What matters is to understand how deeply rooted factors—factors that have shaped mindsets and behaviors over time and created the true wealth of mankind in all its extraordinary diversity—influence development processes positively or negatively.

Cultures and related mindsets and behaviors are very slow to change, and it may be that the globalization process, instead of leading to uniformity, pushes civilizations and nations to intensify their specificities, thereby contributing to a healthy diversity (D'Iribarne and Henry 2003). Cultural features present both strengths and weaknesses, and the policy implications are clear: build on one's natural strengths while being conscious of one's weaknesses.

Table 3.4 *East-West Contrasts in Economic Systems*

	West	*East*
Innovation/industry	Importance of science-based innovations, technology leaders (radical innovations)	Technology- and production-driven innovations, technology followers (FDI, licenses)
Education	Education values sense of individualism, exploration of unknown	Education promotes collective values, imitation
Institutions	Public/private systems regulated by the rule of law, formal contracts	Connection-based systems with informal relationships (*guanxi*)

Source: Aubert 2006.

China illustrates the application of a number of the policy principles outlined above. It has successfully:

- Built on strengths, first by attracting FDI for cheap mass production and manufacturing, and then moving gradually to establish an indigenous R&D capability
- Exploited agents of change, such as local communities, top universities, and high-technology diasporas
- Launched pilot operations such as technology parks along the coast, scaling up those that were effective.

These actions reflect features inherent in Chinese culture: pragmatism, openness, experimentation, adaptation to long-term trends, along with the ability to seize short-term opportunities. On the other hand, some problematic aspects of the culture have been insufficiently corrected: lack of attention to public goods (support of basic research), and governance based on personal relationships (*guanxi*) leading to a neglect of fair competition. These issues clearly affect Chinese growth perspectives in the long term (see chapter 8).

In sum, it appears that development processes are based on an appropriate blending of tradition and modernity, and global and local good practices. This should inspire the building of development strategies, particularly those based on knowledge and innovation.

4

Improving the Economic and Institutional Regime

A good economic and institutional regime is of primary importance for achieving better policy results in the functional knowledge pillars, as well as for getting the most from related investments. Following brief general discussions of governance conditions and the business climate in low- and medium-income countries, this chapter focuses on a few issues essential to any knowledge economy (KE): regulation, finance, and trade. It concludes with comments on local reforms, which in developing countries often are a way forward.

General Considerations of Governance and Business Climate

Advanced economies typically have well-established institutional frameworks based on democracy and free markets. But to favor the development of a knowledge economy the institutional framework must go farther, extending to labor markets (employment flexibility, employability), sophisticated financial markets (venture capital), products and services markets, and effective protection of intellectual property. In other words, the basic economic and institutional framework requires fine tuning from a KE perspective. Of course, even fine tuning can encounter resistance from vested interests or be slowed for other reasons.

In medium-income countries, more than fine tuning is generally needed to adapt the economic and institutional regime to KE needs and opportunities. However, experience shows that when countries are keen to insert themselves into the global economy, important reforms are carried out and vested interests overcome, if only in key sectors of the economy. Among transition economies, China and the countries of Central Europe stand out as examples.

The often mediocre economic and institutional regimes in most low-income countries raise important challenges. The (re)construction of states cannot be accomplished overnight—the effort requires a well-articulated, strategic set of steps focused on very specific problems and taking into account bureaucratic, political, social, and economic interests (box 4.1).

Achieving good governance nationwide through a sweeping program of economic and institutional reforms is a tall order—and successes are rare. The occasional victories have occurred in countries that have made a smooth transition to modernity, without a dismantling of traditional social structures owing to colonization, decolonization, or other historical factors. The case of Botswana, a country that was relatively isolated from the colonization process (Leith 2005), is particularly illuminating in this perspective (box 4.2).

Box 4.1 Improving Governance in Low-Income Countries

If the public sector in African nations is to function well, governments and their partners must enlarge their present focus on organizational, technocratic, and public-management approaches. That is the conclusion of a World Bank study on governance in Africa. Authors Brian Levy and Sahr Kpundeh (2004) urge a broader perspective that incorporates the political dynamics of the country and the institutional rules of the game within which public organizations operate. Some key lessons:

- In reforming state institutions, it is important to find the right fit. Systems of accountability are country-specific, so cookie-cutter approaches, even those deemed "best practice," are doomed to fail if undertaken without first being customized to the circumstances at hand. Efforts to strengthen administrative and accountability systems have to be made to fit each country's constitutional structure and patterns of political, social, and economic interests.
- Building states that are both effective and accountable to their citizens is sometimes a centuries-long process, but small beginnings over time have large effects.
- Comprehensive reforms may not be feasible if the country does not have the bureaucratic and institutional capabilities to implement them. It may be better to focus on modest but viable initiatives for which results are readily observable. For example, if it is not possible to fix the whole government, getting community schools to work may spearhead more reforms down the road.
- Public administrations operate in complex and interdependent systems of bureaucratic, political, social, and economic interests, so efforts to build state capacity must take into account the underlying drivers of political and institutional change. This approach complements a previously prevailing view that traced problems to poor management and endeavored to fix them through reorganization, technical training, and new hardware.
- Corruption stifles economic growth but affects the poor disproportionately by increasing the price they pay for public services and restricting their access to essential services. The roots of corruption lie in dysfunctional state institutions. Anticorruption campaigns can work, but only when pursued in tandem with institutional interventions.

Source: Levy and Kpundeh 2004.

Obstacles to Business Development

The World Bank's 2005 survey of world investment climates identified a set of common obstacles to business formation and growth (World Bank 2005a). Firms perceive political instability, macroeconomic instability, high taxes, corruption, and access to finance as the most severe obstacles to growth and development. But other factors also loom large—among them crime, ineffective legal systems, burdensome administrative regulations, and poor infrastructure (figure 4.1). Other econometric research has revealed that finance, crime, and political instability present the greatest obstacles to firm growth and most directly affect the rate of growth. That is, they are the binding constraints to growth (box 4.3).

Figure 4.1 depicts the "standard" developing country. Although obstacles will vary in importance from country to country, most countries moving toward a knowledge economy will have to consider *all* the items listed as priorities—especially those that the "standard" country considers low priority: skills, labor regulations, transportation, and telecommunications.

Box 4.2 Success in Achieving Significant Economic and Institutional Reforms: The Case of Botswana

With rapid growth, Botswana has become a middle-income country (see box 3.2). Its achievement can be traced in good part to good governance and growth-promoting economic policies.

- The country's minerals policy established mutually profitable arrangements with foreign investors and one of the very few successful international marketing arrangements. Sales of diamonds and other minerals generated substantial rents that might simply have been spent, but the government's long-term development plans channeled funds into investments that promoted both growth and human development while maintaining fiscal discipline.
- Trade policy kept the economy open to competition from imports and maintained access to markets for some important nonmineral exports.
- Money and banking policies, while not always optimal, generally provided stability to the economy and to the financial sector.
- Exchange rate policy ensured stability in exchanges of goods and services and avoided the danger of an overvalued currency, a common phenomenon in states rich in natural resources.
- Fiscal policy, on the whole, has been disciplined. The accumulation of substantial government savings and foreign exchange reserves has enabled the country to ride out downturns in the diamond market, and the careful management of those reserves has generated a significant return to the nation.
- Labor market policies avoided both the extremes of an exorbitant real minimum wage and a bidding war for scarce talent. The government pursed a policy whereby wages and salaries in the private and parastatal sectors generally conformed to those paid to comparable grades of public employees.

Sources: World Bank 2000a; WEF 2003; U.S. Department of State 2003, 2005; IMF 2004; Land 2002; Transparency International 2005.

Figure 4.1 How Do Firms in Developing Countries Rate Various Investment Climate Constraints?

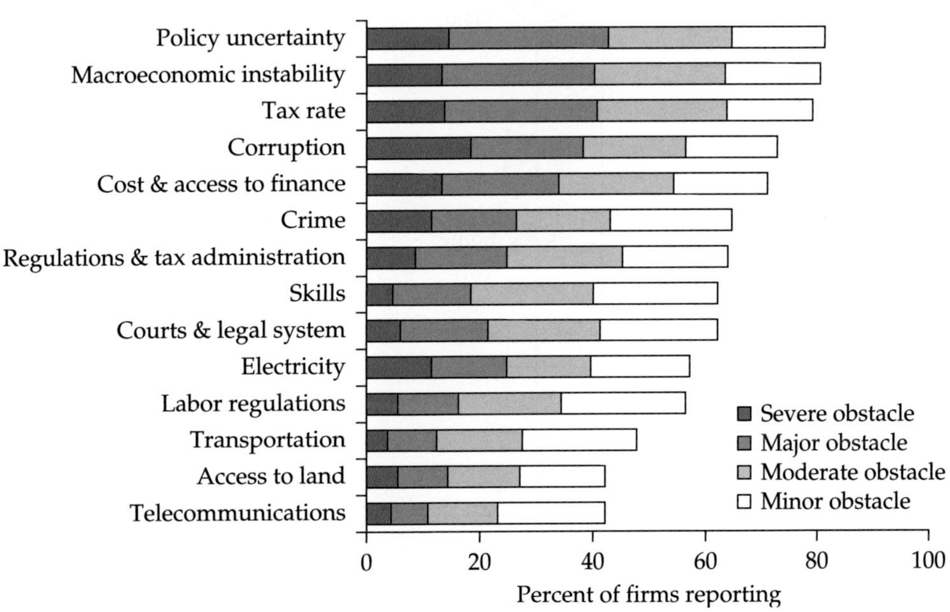

Source: World Bank Investment Climate Surveys.

Box 4.3 *Obstacles to Firm Growth in Developing Countries: What Are the Binding Constraints?*

Firms throughout the world agree on the obstacles to growth, but there is little empirical evidence on the extent to which each actually constrains growth. Using firm-level survey data from 80 countries to marshall evidence on the relative importance of different features of the business environment, Ayyagari, Demirguc-Kunt, and Maksimovic (2006) examine factors such as taxes and regulations, judicial efficiency, infrastructure weaknesses, and financing issues. They reach some interesting conclusions. According to their econometric research, only access to finance, crime, and political instability are *binding constraints* on the growth of firms, with finance the most important. Their findings have important implications for the planning of reforms. Maintaining political stability, controlling crime, and relaxing the financial constraints that weigh on firms, while tackling other bottlenecks, could be the most effective route to firm growth.

Source: Ayyagari, Demirguc-Kunt, and Maksimovic 2006.

Investment climate assessments conducted under World Bank auspices help developing countries identify other obstacles to investment, growth, innovation, and initiative. The obstacles are diverse. They include government authorizations of various types, government procurement practices, technical norms and standards, customs procedures, industry-university relations, finance and banking practices, intellectual and other property rights, and the competitive climate, which is particularly important for stimulating innovation (figure 4.2). Such obstacles are not fundamen-

Figure 4.2 More Competition Means More Innovation

Degree of competitive pressure in economy

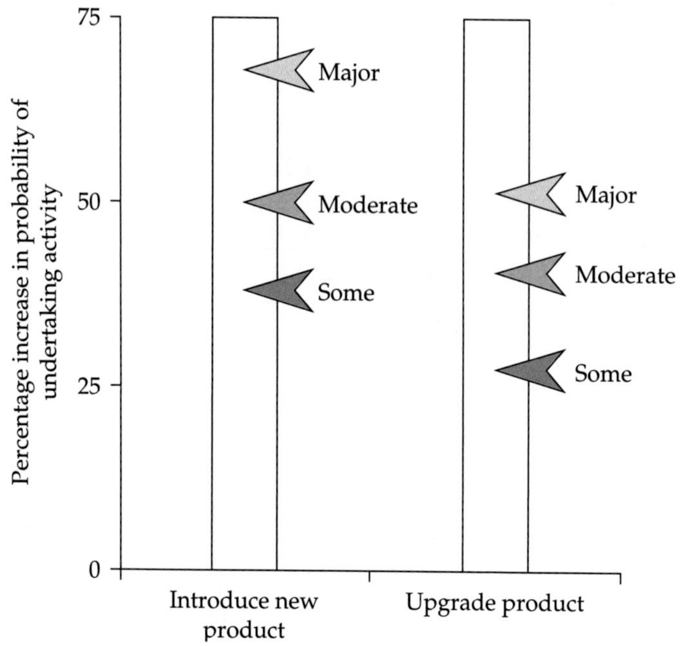

Source: World Bank 2005a.

tally different from those encountered in the developed world, but they can be much more difficult to address, often because of the absence of an efficient judiciary system. Of course it is not enough to identify obstacles. Audits and assessments should be followed by sustained efforts to ensure that the problems are resolved.

Detailed analysis at the sector level often is needed to provide policy makers with solid policy advice. Such analyses have been developed by specialized organizations such as the McKinsey Global Institute (Palmade 2005), which conducted studies of representative samples of industries in 20 countries between 1990 and 2003 to identify the causal links between factors in the firms' external environment and their behavior. Analyzing the competitive dynamics in key sectors can yield accurate growth diagnostics and help identify binding constraints at the sector level. It would be useful to enlarge such audits to include KE-related issues.

Regulation

The tendency to overregulate is found in many developing countries as shown in figure 4.3, which depicts the number of days required for various business procedures at different levels of economic development. But some countries have found ways to adapt their regulatory framework to remove bureaucratic and other obstacles that sour the business climate, discourage enterprise creation, and frustrate management (table 4.1). When the good practices discovered by policy makers in one country inspire those in another, the imitators must tailor the borrowed practices to their own situation.

If a government is to succeed at reform, action must be focused in the areas that appear likely to have the greatest systemic effects, and a strong power structure is needed—both to facilitate identification of needed changes and to enforce them. Mexico's program of deregulation, conceived in the late 1980s, was successfully implemented in the 1990s by a strong power structure (box 4.4).

Figure 4.3 Poor Countries Regulate Business the Most

Number of days required for each procedure

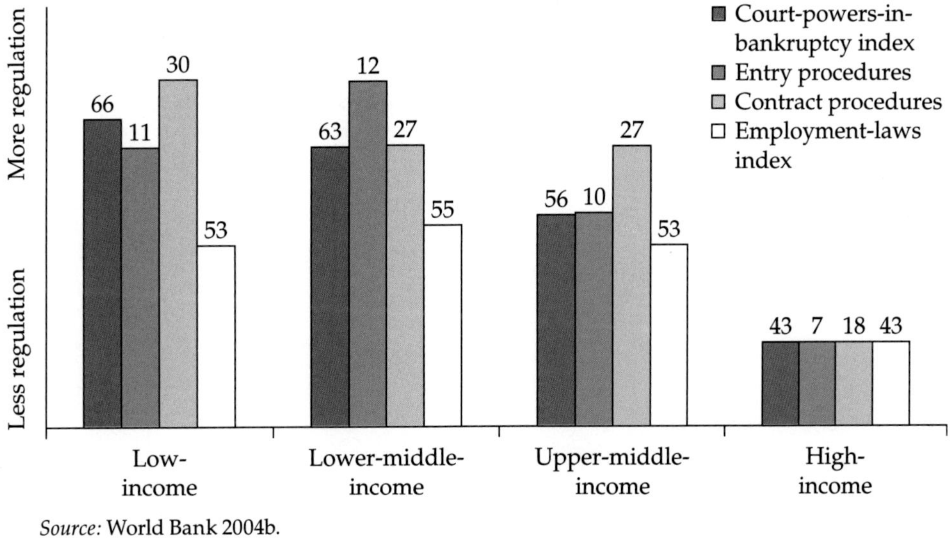

Source: World Bank 2004b.

Table 4.1 Some Simple Solutions to Overregulation

Starting a business Registration as an administrative process *Canada, Chile, Italy, Serbia and Montenegro* Use of single identification number *Belgium, Estonia, Morocco, Turkey* No minimum capital requirement *Botswana, Ireland, Tanzania, Thailand* Electronic application possible *Latvia, Moldova, Sweden, Vietnam*	**Enforcing contracts** Summary proceedings for debt collection *Bosnia-Herzegovina, Finland, Philippines* Case management in court *India, Malaysia, Slovakia, United States* Appeals limited *Botswana, Chile, Estonia, Greece* Enforcement out of court *Hungary, Ireland, Netherlands, Sweden*
Hiring and firing workers Long duration of fixed term contracts *Austria, Costa Rica, Denmark, Malaysia* Apprentice wages for young workers *Chile, Ecuador, Finland, Tunisia* Redundancy as grounds for dismissal *Armenia, Botswana, Lebanon, Russian Federation* Moderate severance pay for redundancy *Finland, Madagascar, Namibia, Uruguay*	**Protecting investors** Derivative suits allowed *Chile, Czech Republic, Republic of Korea, Norway* Institutional investors active *Chile, Republic of Korea, United Kingdom, United States* Disclosure of family and indirect ownership *Denmark, Sweden, Thailand, Tunisia* Public access to ownership and financial data *Germany, Poland, South Africa*
Registering property Consolidate procedures at the registry *Lithuania, Norway, Thailand* Unify link between cadastre and property *Australia, Netherlands, Slovakia* Make the registry electronic *Italy, New Zealand, Singapore* Complete the cadastre *Austria, Czech Republic, Denmark, Ireland*	**Closing a business** Foreclosure focus in poor countries *Armenia, Kenya, Nepal, Paraguay* Specialized expertise in courts *Colombia, India, Latvia, Tanzania* Appeals limited *Australia, Estonia, Mexico, Romania* Administrators paid for maximizing value *Denmark, Jordan, Japan, Malaysia*

Source: World Bank 2004b.

Box 4.4 Successful Deregulation in Mexico

In 1988 the president of Mexico appointed a deregulation czar, who was to report directly to the president and his economic council of ministers every month. Every business in Mexico, large or small, was promised equal access to the czar's office to complain about burdens imposed by government rules and regulations. When the office received a complaint, it was obliged to find out why the rule existed, how it interacted with other regulations, and whether it should remain in effect. The office operated under a strict timetable: if it did not act to maintain or revise a disputed rule within 45 days, the rule was automatically annulled.

The work of the deregulation czar during his first four years is widely credited with accelerating Mexico's larger reform program. It provided struggling private businessmen and women with an effective, responsive champion at the highest level of government. Several factors contributed to this success:

- The czar had unequivocal presidential support; this signaled to bureaucrats and citizens alike the need to comply with his decisions.
- The czar's decisions could be overruled only at the highest level of government.
- Tough penalties were imposed on officials who failed to implement the czar's rulings.
- The czar operated under a strict time limit, which ensured quick and visible results.
- The czar's staff understood the economic consequences of regulations, the complicated interactions within the regulatory field, and their administrative requirements.
- The czar won credibility with officials and the public by giving a fair hearing to both the powerless and the influential and by establishing a record of impartiality.

Source: Authors.

Finance

A knowledge economy cannot be built without finance. The sums need not be large. For small entrepreneurial projects in developing countries, funding needs may be relatively small, and microfinance mechanisms may suffice. Spreading rapidly throughout the world, following the pioneering initiative of the Bangladesh Grameen Bank, microfinance relies on the social responsibility of borrowers belonging to a narrow group to ensure repayment.[1] Small loans from microfinance institutions help entrepreneurs set up the new small businesses that are important disseminators of new technologies at a very basic level and can help improve productivity in the informal economy. Widely reported examples include cell phone shops and Internet cafes. They also support creation of small niche activities based on specialized skills, such as handicrafts.

1. Social responsibility works through the commitment of groups of persons (parents, friends, etc.) who guarantee the entrepreneur's loan, ensuring repayment in case the borrower defaults. Examples of successful microfinance initiatives include Credife in Ecuador, MiBanco in Peru, CrediAmigo in Brazil, SEWA in India, Bank Rakyat in Indonesia, and numerous initiatives in Arab countries. In Morocco, the number of borrowers increased from 50,000 to 600,000 from 2000 to 2005.

Other entrepreneurial projects require a significant amount of development capital. Indeed, a broad range of financial services is required to support growth and entrepreneurship in knowledge-based economies in the developing world, as elsewhere. But in the developing world the difference between traditional banking and corporate finance, on the one hand, and new-venture financing, on the other, is not fully appreciated or exploited. Unlike traditional financial services, even those designed for small and medium enterprises (SMEs), venture capitalists provide seed money and start-up capital, as well as growth capital.

In the countries of the Middle East and North Africa, for example, early efforts to develop SME financing were limited to encouraging banks to extend existing services to the SME sector (European Commission 2005). Later, a new form of funds, known as "general funds," was put in place. They were general in the sense that they would consider investing in almost any sector and at almost any stage of financing. But they faced an uphill battle because company owners, unfamiliar with finance, had difficulty understanding their potential importance. And many owners were reluctant to take advice from relatively young fund managers. A second generation of funds shows a trend toward market segmentation. Funds now target specific sectors such as tourism, agro-food, and information technology. Faced with the specialization of funds for financing enterprises, some governments, in Morocco for example, have established regional investment centers that cater to the needs of foreign investors as well as to those of local entrepreneurs and business people who might require access to finance for growth or expansion.

Financing in an emerging KE economy requires solid and moderately sophisticated capital markets to allocate credit when and where needed by promising firms. Dynamic capital markets can also be useful as platforms for second-tier funding mechanisms for venture capital (so-called VC funds of funds), which are now common in advanced countries. Solid capital markets help in refinancing and diversifying resources for "funds of funds."

In many areas the financial liberalization policies pursued in the 1990s have not yielded the expected results. In some cases they have led to crisis, as in the Russian Federation, Asia, and Latin America. Overall, "the development of the framework for capital markets—such as reasonable information, legal and judicial treatment of bankruptcy, conduct rules for market participants, etc.—was slow" (Zagha and Nankani 2006). Credit allocation continues to favor the public sector, well-connected individuals, financial-industrial conglomerates, and other long-time clients of state banks. This adherence to old ways can stall KE efforts by choking off funding for new, innovative, and growing firms.

The availability and use of venture capital in most developing countries remain embryonic. In some middle-income countries, however, a growing contingent of specialized firms and individuals—business angels—mentor budding entrepreneurs and inventors and contribute some start-up capital, modeling their actions on their counterparts in more advanced economies.

Expatriates and diaspora networks can be an important source of venture capital and related expertise. Migrants' remittances are a massive source of funds for many developing countries, dwarfing flows of foreign aid by a factor of ten or more. But most remittances are channeled quickly into consumption or invested in traditional assets, especially real estate. Redirecting a small percentage of these sums to innovative ventures could make a considerable contribution to KE financing. The channel-

Box 4.5 Creating a Venture Capital Industry: The Network of Overseas Chinese

When the Taiwanese government decided to promote a venture capital (VC) industry in the beginning of the 1980s, it had neither blueprint nor capability. Many were opposed to the idea because the concept of venture capital was foreign to traditional Taiwanese practice, in which family members retain close control of a business's financial affairs. Moreover, entrenched interests sought to maintain the status quo. Through a process of intense interaction with the Taiwanese diaspora in Silicon Valley, however, new institutions such as the Seed Fund (with an initial capital allocation of NT$800 million, later increased by additional NT$1.6 billion) matched the contributions of private venture capitalists.

Two American-style venture funds, H&Q Asia Pacific and Walden International Investment Group, were created in the mid-1980s, managed by U.S.-educated overseas Chinese invited to relocate to Taiwan (China). Once the first funds proved successful, domestic information technology firms created their own VC funds. And as those started to pay off, even the conservative family groups began to invest in VC funds and IT businesses.

The lesson? A network consisting initially of dynamic and forward-looking members of the Taiwanese government and leading overseas Chinese engineers in Silicon Valley was central to the emergence of a modern VC industry in a society dominated by conservative and risk-averse business groups. While the network did not have a blueprint, it had a role model (Silicon Valley) and a clear idea of what to do next. By defining each step along the road, the network expanded, eventually incorporating skeptics and opponents.

Source: Authors, based on Saxenian 2000.

ing of such resources requires well-designed mechanisms and learning processes, as the experience of Taiwan (China) in the early 1980s demonstrates (box 4.5).

Appropriate tax incentives can encourage the mobilization of development and venture capital. However, their effects depend significantly on the overall organization and structure of the tax regime. Moreover, they require sophisticated institutional arrangements that are free from corruption, but often these are not in place in developing countries. As a result, some countries have chosen to establish a flat tax with (relatively) modest rates of around 20 percent applied to both corporations and individuals. This has proved an efficient stimulus to business creation and development in several countries (Ireland, Estonia, and Slovakia, for example), in part because they have simplified tax procedures for both payers and collectors.

Trade

Trade is essential for generating the foreign exchange required to import needed technologies, production processes, and capital equipment. Once obtained through trade, such knowledge goods and services boost the productivity and competitiveness of domestic producers. But trade contributes to productivity growth in another way as well—it can increase competition in ways that stimulate industries to become more efficient. Therefore, policies that shelter the economy from the world market block beneficial spillover benefits and the dynamic gains to be had from integration (box 4.6).

Most countries can obtain immediate benefits by unilaterally dismantling barriers to trade and investment. But trade performance, and the gains derived from trade, may

Box 4.6 What the 1990s Revealed about Trade Reforms

- **Openness to trade has been a central element of successful growth strategies.** In all countries that have enjoyed sustained growth, the share of trade in GDP has increased, and trade barriers have been reduced.
- **Trade is an opportunity, not a guarantee.** While trade reforms can help accelerate integration into the world economy and strengthen an effective growth strategy, they cannot ensure its success.
- There are **many different ways to open an economy.** The challenge for policy makers is to know the country's specificities well enough to be able to identify which reforms best suit their country's political economy, institutional constraints, and initial conditions. Because these vary from country to country, it is not surprising that one finds a striking heterogeneity in countries' experiences of the timing and pace of reforms. Specific political economy considerations, such as adjustment costs, need to be taken into account at the design stage if reforms are to be sustainable. China, for example, created some new profit opportunities while leaving much of the old order undisturbed. As a result, there were no identifiable losers. Few vested interests opposed the special economic zones because these were set up outside the scope of central planning and did not disrupt existing production and allocation. This approach also maximized political support for the reforms as the number of winners rose over time.
- One element, however, is common to almost all success stories of the 1990s: despite the diversity of approaches to trade reform, **all successful liberalizations have promoted export growth,** explicitly or implicitly. Incentives to exporters ensure that selling on international markets is as attractive as domestic sales.
- The **distribution effects of trade liberalization are diverse,** and not always pro-poor. Despite expected gains for the economy in the longer term, trade reform generates both winners and losers in the short run. The poor are more likely to share in the gains from globalization if complementary policies are in place. A key challenge for policy makers is to design targeted social safety nets and complementary policies to ease the transition from contracting to expanding sectors.

Source: Adapted from Zagha and Nankani 2006.

also depend on the actions of trading partners. If developing countries are to benefit fully from their comparative advantages, expand their trade, and move into higher-value-added goods and services, barriers to their exports must be removed or substantially reduced. Achieving the necessary changes requires collective action in the form of international negotiations. Four types of trade impediment warrant particular attention from a KE perspective: tariff peaks, tariff escalation, contingent-protection measures (antidumping and safeguard rules), and product standards. All four are currently the object of negotiations at the bilateral, regional, and multilateral levels.

For KE-based growth to take off, trade-related support systems are essential. Among the most important are:

- Infrastructure with effective port, airport, refrigeration, container, and storage systems, as well as related trade logistics
- Custom controls, which can now be made better and more efficient by electronic systems
- Certification, such as phytosanitary controls and International Standards Organization (ISO) systems.

Beyond these trade-related measures, other actions are needed to promote and protect the various intangible assets that are crucial in efforts to develop a knowledge economy. Examples include networks of science attachés (to scout for needed technologies), scientific resources, and business partners; systems of visas and work permits to facilitate travel and immigration of knowledge workers and specialists; and bodies to coordinate participation in trade and technology fairs.

Action is also needed to promote foreign direct investment (FDI). Attracting foreign investors requires not only measures such as tax incentives and good infrastructure, but also promotional campaigns to draw attention to those advantages. Experience shows that efficient FDI policies require the establishment of powerful agencies to coordinate the many different instruments needed to attract and support foreign investors.

Moreover it is essential to stimulate by all appropriate means the transfer of competencies and technologies between foreign firms and local industry. Few countries have succeeded fully in doing so. The Irish Linkages Program described in chapter 5 has done better than most, however, and can show developing countries how to take better advantage of the foreign firms they host.

Clusters and Related Measures

Clustering is generally defined as a process in which firms and other actors locate near one another. Clusters are generally driven by the private sector but may be set up either by the government or spontaneously. Participants in the cluster cooperate around a certain functional niche, forging close links and working alliances to improve their collective competitiveness in the anticipation of mutual benefits. The categories of cluster participants are illustrated in figure 4.4.

Figure 4.4 Different Cluster Actors

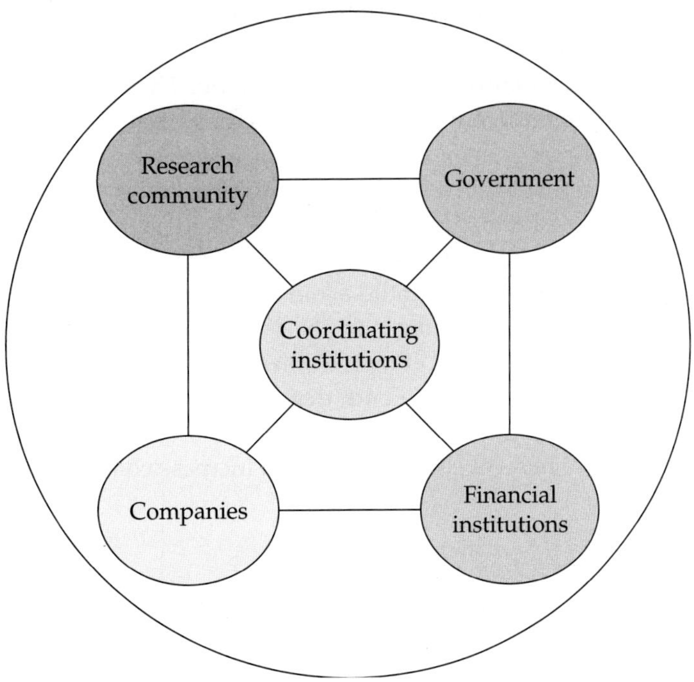

Source: Adapted from Andersson and others 2004.

> *Box 4.7 Public Policies That Encourage Cluster Development*
>
> Public policy of various types can usefully encourage clusters.
>
> - ***Brokering policies*** enable value-enhancing dialogue and collaboration beyond what would be achieved without the broker. Such policies strengthen the framework for dialogue and cooperation among the stakeholders involved in clusters. Brokering policies should not favor individual players; they should be facilitative and noninterventionist.
> - ***Demand-side policies,*** such as public procurement, should seek to increase openness to new ideas and innovative solutions. This policy instrument should be used with caution.
> - ***Training policies,*** such as vocational training programs, may be targeted at upgrading the skills and competencies that are essential for effective clustering of small and medium enterprises.
> - ***Promotion of international linkages,*** such as the elimination of trade barriers and strengthening of transport and communication systems, along with the harmonization of market regulations, can enhance the interplay of foreign and domestic actors.
> - ***Framework policies*** for macroeconomic stability, well-functioning product markets, factor markets, education, and good governance should have effective and consistent rules for inter-actor transactions.
>
> *Source:* Adapted from Andersson and others 2004.

Because clusters can yield a range of benefits, such as improved opportunities for innovation, greater business formation, and higher productivity, public policy has been increasingly keen on setting up and promoting cluster initiatives. Some of the policies that may usefully promote a cluster are described in box 4.7. Although government actions should facilitate rather than intervene, the approach to cluster policies should be comprehensive.

Technology parks can encourage cluster formation. The establishment of such parks, which are different from industrial zones, has been a favored method of supporting new technology-based companies in a number of developing countries. Experience in industrialized countries has shown, however, that success is far from guaranteed and that policymakers should proceed cautiously and gradually. A good approach is to build parks close to universities and link them to dynamic business communities. It is important to avoid the use of financial incentives, particularly tax incentives, that give firms in the parks (and their personnel) advantages that may be perceived as unfair by the other parts of the business or academic communities. There is a tendency in both developed and developing countries to develop parks around specific technologies, such as biotechnology or ICT, in order to concentrate a critical mass of competencies and increase chances of success.

Local Reforms

Nationwide reforms of the business environment in developing countries are almost as rare as reforms of the entire governance framework. In fact the two are complementary and generally take place only when sweeping political changes occur, such as those in Eastern Europe and some countries of the former Soviet Union

Box 4.8 China's Special Economic Zones

Primarily geared to exporting processed goods, SEZs are foreign-oriented areas that integrate science and industry with trade and benefit from preferential policies and special managerial systems. Indeed, the word "special" in SEZ refers to special economic systems and policies. China's central government gave SEZs policies and flexible measures such as special tax incentives to attract foreign investments (significant reduction or elimination of customs duties and income tax), as well as greater independence for international trade, at a time when the rest of the economy was predominantly state-controlled.

The first SEZs were established in the early 1980s in Shenzhen, Zhuhai, and Shantou in Guangdong Province and Xiamen in Fujian Province. In 1984, China opened 14 coastal cities to overseas investment: Dalian, Qinhuangdao, Tianjin, Yantai, Qingdao, Lianyungang, Nantong, Shanghai, Ningbo, Wenzhou, Fuzhou, Guangzhou, Zhanjiang and Beihai. 1988 saw the designation of the largest SEZ so far, Hainan Island. In June 1990 the Chinese government opened the Pudong New Area in Shanghai, the "dragon head" of SEZs, to overseas investment, as well as additional cities along the Yangtze River valley. Since 1992, the State Council opened a number of border cities, and in addition, opened all the capital cities of inland provinces and autonomous regions.

In addition to these SEZs, 15 free trade zones, 32 state-level economic and technological development zones, and 53 new- and high-technology industrial development zones have been established in large and medium-sized cities.

These areas have played a dual role: (a) that of a "door" to the world through which products leave and foreign exchange and advanced technologies enter, and (b) that of "spearhead" in accelerating local and regional economic development.

Shanghai and Shenzhen are certainly the best-known SEZs, although Tianjin's TEDA has done particularly well also.

Source: Adapted from EIU 2005.

in the 1990s. Wholesale reforms following the collapse of communism brought a vastly improved business climate in Estonia (1991), Slovakia (2001), and Georgia (2005). Such occurrences, however, are infrequent. This is why efficient action to ignite and sustain growth often requires more selective approaches focused on specific territorial areas, as China has done to spectacular effect (box 4.8). As discussed in chapter 3, such locally focused reforms are often an efficient way to initiate the "bootstrapping" process that can lead to KE-based growth.

In any event, decentralization and power devolution are of great importance in the developing world, particularly for designing and implementing knowledge-related policies. It is essential to mobilize and empower local communities and regions, especially where the nation state is not yet firmly established, as in some low-income countries. The mobilization process requires clear guidance from the center to ensure coherence in the actions taken. The necessary financial resources should be available as appropriate, through tax levies or redistribution mechanisms, a point discussed elsewhere in this volume.

5

Fostering Innovation

Innovation policy, although fashionable, is often misunderstood; it is not an appendix to science and technology policy, as often presented. Innovation—the application of knowledge of all types to achieve desired social and economic outcomes—is broader than science and technology, often combining technical, organizational, and other sorts of changes (World Bank 2007).

This chapter begins with some general considerations of the notion of innovation, particularly in the context of developing countries, and of the institutions and instruments commonly used to promote it. Next, we discuss the two major functions of innovation policy. These are support for innovators and the development of research and technology infrastructure. The two go together because innovation results when entrepreneurs cross paths with scientific and technological assets. We move on next to two major issues facing developing countries: the diffusion of basic technologies in poor areas and the promotion of export industries. Finally, we offer some insights on issues relating to intellectual property rights (IPRs), which are the subject of important global debates.

General Considerations

Understanding Innovation

In the context of developing countries, innovation is the design, development, and diffusion of a technology or practice that is new to the society in question (Aubert 2005b).[1]

Innovation occurs at three levels:

- Local improvements are made by adopting available technologies to satisfy basic needs or to upgrade products or services.
- Competitive industries develop through adaptation of technologies initially produced in or by developed countries.
- Ultimately, new innovations of global significance are developed.

Innovation policies in developing countries must begin by building up an appropriate technical culture and establishing incentives to support and stimulate entrepreneurship. A general starting point is to reverse the usual model, of questionable

1. As a complementary definition, innovation is the process by which organizations and enterprises "master and implement the design and production of goods and services that are new to them, irrespective of whether they are new to their competitors, their country, or the world" (Mytelka 2000).

relevance even in the industrialized countries, in which innovation begins with, and derives from, research. In low-income countries, it is far more useful to look at ways in which industrial technology is gradually introduced and upgraded at both the micro and macro levels (figure 5.1). Technological performance, upon which economic performance depends, is shaped first by basic operational and craft skills. The next stage is product design and engineering. Research and development (R&D) activities needed to develop new technologies come much later, after firms have acquired sufficient intellectual and financial capital. The same sequence applies to the economy as a whole. Innovation helps generate the resources needed for research.

Particularly appropriate for developing countries are innovations based on good use of local assets exploited by local communities. An exemplary model, "brand agriculture," comes from Japan, an industrialized country, and is being applied by rural communities in developing countries. This model offers a way to shift from subsistence agriculture to more competitive agriculture (box 5.1).

Figure 5.1 Structure of Industrial Technology

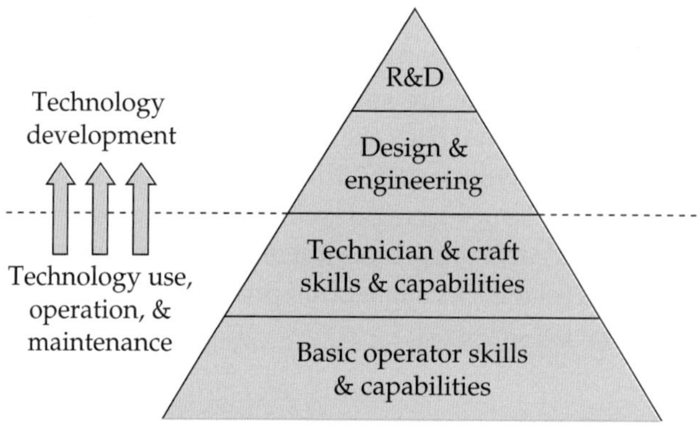

Source: Adapted from Intarakumnerd and others 2002.

Tapping into Global Knowledge and Technology

Complementing local assets as sources for innovation are global knowledge and technology, which can be tapped either formally or informally. Countries that are still behind the technological frontier are likely to derive greater increases in productivity and welfare by acquiring existing knowledge than by pursuing homegrown R&D to develop new knowledge, which is generally more difficult and risky and requires much more technological capability than obtaining and adapting existing knowledge.

Openness to trade provides firms and households with access to world markets for goods, services, and ideas. Access brings lower prices, better quality, and greater variety of consumption goods, while encouraging specialization of economic activity in areas where countries enjoy a comparative advantage. Openness to trade also tends to attract investment and to foster higher productivity of domestic industries as domestic producers emulate foreign competition and mine foreign producers for knowledge (see chapter 4). Trade and foreign direct investment (FDI) are particularly important for generating the spillover benefits associated with learning through diffusion and absorption of technology.

Box 5.1 *Pride of Place: Innovating Successfully in Rural Areas*

Challenging the common literature of economic growth and development, which tends to assign the agrarian sector to the backwaters, Japan is putting agricultural and rural development at the forefront through a community-based strategy designed to stimulate innovation and human capital accumulation.

The concept, known as "brand agriculture," is a general strategy for community-based rural development that identifies, cultivates, and exploits local resources (natural, historical, and cultural and human resources, including women and elderly people) for the development of products and services unique to a certain village or geographical area. The process feeds the engine of sustained development of a greater variety of these products and *gradually establishes these local brands in increasingly larger markets.* In the process, the community accumulates technical skills, know-how, and practical knowledge learned by inference through experience (tacit knowledge).

Two examples of Japanese brand agriculture that came about as spontaneous grassroots movements for rural development are presented: One Village One Product (OVOP), and "roadside stations" (*michino eki*). Both consider agriculture broadly to include forestry, fishery, and stock-raising, as well as industries and activities based on local resources such as local food processing, crafts, restaurants, and tourism. Through increasingly sophisticated marketing, unique local products have been sold in wider markets, often establishing distinctive regional brand names identifying the local manufacturers of the products.

A roadside station is operated and managed by the local community under the guidance of the Infrastructure and Transport Ministry. A typical station sells a large set of unique local products developed and produced either within the vicinity or in the surrounding rural communities. Hence, local residents (in particular women and elderly people) gain opportunities to increase their income and entrepreneurial experience. The station also serves as a venue for the provision of a wide variety of public services to the local community, such as sanitation, health care, education and training, and cultural activities. Similar projects are now being implemented in Thailand, India, China, Turkey, Kenya, the Republic of Yemen, and Mexico.

The author underlines the importance of human capital and basic services to support the branding strategy—including electricity, water, sanitation, transportation, telecommunications administration, management, financing, marketing, technical, and R&D institutions.

Source: Fujita 2006.

Other means by which technology is transferred include licensing; technical assistance; importation of capital goods, components, and products containing technology; copying and reverse engineering; foreign study; technical information in printed or electronic form (including that available over the Internet); and twinning and training arrangements. Almost by definition, proprietary technology is sold or transferred on a contractual basis, but it may leak out, depending on the strength of a country's IPR regime and the sophistication of the users of the technology. A great deal of relevant technology is in the public domain—or owned by governments that could place it in the public domain. The key issue then is how to ensure that freely available technology reaches those who need it.

One of the main actors in the creation and dissemination of applied knowledge is the multinational corporation. The R&D budget of a large multinational is greater than the total R&D expenditures of all but the very largest of developing countries. For example, General Motors spent $7.9 billion on R&D in 1998, more than India spent in the same year. Developing countries can significantly benefit from the investments of multinationals if they can become integrated, either horizontally or vertically, into a global production chain. The Irish National Linkages program provides an excellent example of how this can be done (box 5.2).

A country's diaspora can be a significant source of knowledge if well used. International mobility of talent and its most visible manifestation, brain drain (usually defined as the migration of human capital from less to more developed economies), is an important and hotly debated issue. But talented expatriates in technical, managerial, and creative fields can contribute mightily to their countries of origin, as the diasporas of China, India, Ireland, Scotland, and Taiwan (China) have shown. Those contributions are further discussed in chapter 7.

The Institutions and Instruments of Innovation Policy

Finland, considered by many experts to be the world's most competitive nation and universally recognized as having the most efficient innovation policy, offers a good example of the power of government to promote innovation (box 5.3). Three aspects

Box 5.2 *Ireland's National Linkage Program, 1987–92*

To increase inflows of foreign direct investment (FDI) and leverage technology to develop an indigenous technological capability, Ireland's Industrial Development Authority designed and implemented the National Linkage Program in 1987.

The three main stakeholders were the government, multinational corporations (MNCs), and small and medium enterprises (SMEs). The government provided the political imperative and oversaw the eight state agencies that were to help SMEs become suppliers to MNCs operating in Ireland.

Electronics was the first target of the program. The sector was large, dynamic, and had a propensity to source locally. Through the Federation of Electronic Industries, firms already operating in Ireland helped meet the program's costs in the first two years, responding positively to appeals from executives of the National Linkage Program and government ministers. Those MNCs and new investors in Ireland were quickly introduced to local sourcing opportunities and asked to supplement the technical assistance being provided by state technical agencies.

SMEs seeking to participate in the program underwent a rigorous evaluation of their existing and potential capabilities against perceived supply opportunities. The assessment also included a detailed examination of the firm's financial management.

Because eight agencies were involved in the program, IDA was able to field a well-balanced and multifaceted team of experts in management, business development, technical issues, accounting, and banking. The team quickly developed close relationships with key MNCs and, in close cooperation with MNCs, assessed, selected, and guided new SME suppliers.

The program paid off. From 1987 to 1992, locally sourced electronics materials increased from 9 to 19 percent of MNC purchases. Of a total population of 900 MNCs in Ireland in 1992, some 200 proved to be effective participants in the program. The core group of 83 SME suppliers participating in the program dramatically outperformed similar companies, a result attributable to the selection process, intensive support, and interaction with demanding customers that forced the supply companies into a competitive mode. Over the period, the core group posted average sales growth of 83 percent; an average productivity improvement of 36 percent; and average employment growth of 33 percent.

Source: Authors.

Box 5.3 *The Institutional Features of Finland's Innovation Policy*

Three features of Finland's innovation policy deserve mention. First, the policy is based on a very clear understanding of the public sector's role in the innovation process, particularly with regard to funding. Public institutions and funding mechanisms play a key role in the early stages of the innovation process (basic and applied research), whereas the private sector becomes more involved at the stages of development and commercialization. TEKES is an agency formed by the Finnish government in the late 1980s to stimulate innovative R&D, notably by inducing universities and industries to work together on specific projects as a condition for receiving matching funds.

The Innovation Environment in Finland: Resources and Funding

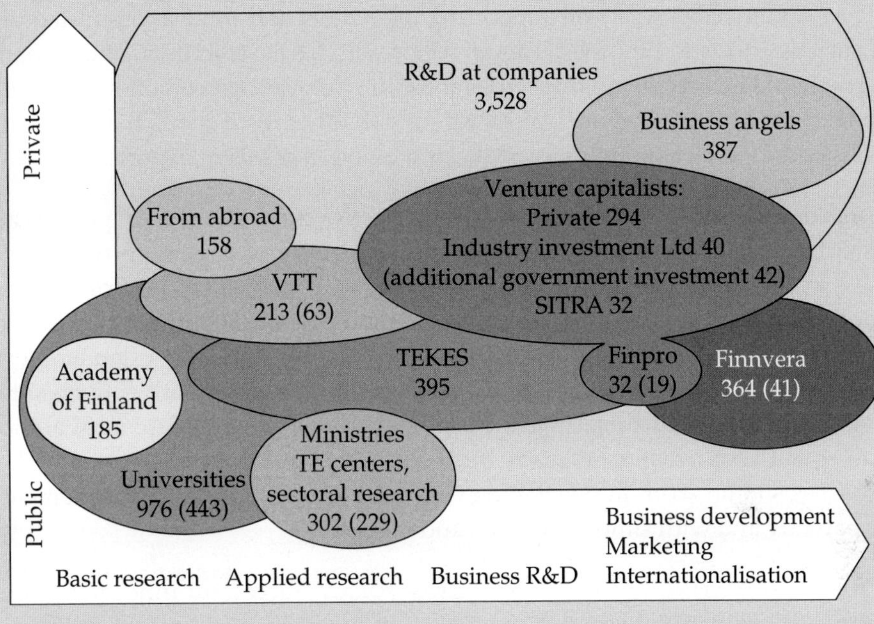

Source: Dahlman, Routti, and Yla-Anttila 2006.

Note: The numbers represent total R&D of each organization in millions of euros in 2004. Funding from the state budget appears in parentheses. R&D by TEKES and the Academy of Finland is funded entirely by the state budget.

Second, Finland's innovation policy has always had a vibrant regional and local dimension. National agencies have regional branches, which enlist local authorities (municipalities, provinces) in pilot programs. Local technology forums convene firms, universities, and other actors to identify opportunities to develop internationally competitive clusters.

Third, innovation policy is coordinated at the national level by a Science and Technology Council chaired by the prime minister. The council brings together key ministers, including the minister of finance, who then mobilize their departments (education, science, industry, trade, etc.) as needed. Representatives of the business community, trade unions, and civil society are also active in the council. Innovation policy is truly interdepartmental, cutting across institutional barriers, a necessary condition of successful government action in this domain.

Source: Adapted from Dahlman, Routti, and Yla-Anttila 2006.

of that power are considered here: (a) government support from an efficient agency throughout the innovation process; (b) attention to the local and regional dimension of innovation; and (c) the key role of innovation policy, clearly understood and implemented as an interdepartmental and cross-sectoral policy.

Appropriate support from government. The first of these points is relevant to the continuing debate over the role and involvement of government in industrial and technological development: governments clearly have a role in supporting "public goods" (notably basic research and risky investments) that benefit the overall economy, but they should not interfere in actions or decisions that are better guided by markets and the working of competition. Moreover, it is crucial that government support should be delivered by an organization that is well-funded, flexible in its modes of action, and focused solely on supporting innovators and innovation-related projects, such as efforts to link the business sector with R&D structures. Those characteristics should be kept in mind when establishing innovation-promotion agencies in national contexts that often are overcrowded with other institutions that have other responsibilities, such as support for SMEs or promotion of FDI or exports.

The local and regional dimension of innovation. The second point is also crucial. Experience shows that innovation flourishes in well-defined regions that possess a concentration of talent, energy, and vision; it is essential for firms' basis of support to be nearby, just as it is important to mobilize local communities to create strong ownership. In developed countries, the mechanisms most often employed include the establishment of local branches of central agencies with enough autonomy to select and fund projects and matching funds that stimulate local authorities to spend resources on infrastructure and other innovation programs. These approaches should work in developing countries as well, if they are implemented transparently and with simple methods and means.

The coordinating role of innovation policy. Successful innovation policy requires a coordinating body that operates across government sectors and departments and involves all key actors, including the minister of finance. Such a body, acting as a catalyst of visions and energies, needs little funding but should have some resources at its disposal to commission consultations and studies, launch promotional and awareness campaigns, conduct technological foresight exercises, and perform institutional audits. It should also have sufficient legitimacy and clout, if not the legal authority, to set priorities and budgets.

Innovation appears to result from the interaction of people who have different competences and assets but converging interests. Therefore, so-called bridging mechanisms are a key instrument of innovation policy. Such mechanisms include matching funds to induce the private sector to collaborate with R&D structures; funding of R&D structures conditioned on collaboration with industry; incubators to provide support to entrepreneurs; extension and productivity centers to facilitate technological information, assistance, and training; and venture capital schemes to mobilize the financial sector.

Stimulating and Supporting Innovators

To create a climate conducive to innovation one must first understand the nature and composition of the domestic enterprise sector. In most developing countries that sector is composed largely of microenterprises that operate in the informal economy with little or no technological competence. A smaller segment consists of SMEs with minimal technological capabilities, while an even smaller segment comprises technology-competent enterprises. Usually only a very small number of enterprises perform significant R&D.

Supporting Enterprise Upgrading

Policy instruments must be adapted to the different types of enterprises, addressing their technical, commercial, and legal needs. Related support needs to be embedded in broad actions aimed at upgrading the overall management of enterprises. Table 5.1 summarizes the policy instruments that can be used to meet these different requirements.

Financial Incentives and Related Support Structures

Governments in the industrialized world have developed a variety of mechanisms to meet the needs of SMEs for innovation-related technical assistance, research, and training. Those mechanisms, described in box 5.4, can be used as models for developing-country policies, although weak administrative and financial structures may require precautions in adapting them.

Box 5.4 Innovation Policy Schemes in OECD Countries

The experience of the industrialized countries in supporting the innovation-related efforts of small and medium enterprises (SMEs) can be very useful for developing countries.

Dedicated financing mechanisms have been a key tool in promoting innovation among SMEs. Most such mechanisms take two forms: (a) provision of seed money in the form of grants to develop ideas, generally up to the prototype stage; and (b) development money for the phase leading to technical realization and market tests, either in the form of matching funds (to match equivalent resources provided by the innovator or another interested party) or subsidies reimbursable in case of success. Further measures have been taken to facilitate the provision of venture capital through public venture capital funds or regional development companies. Another key instrument has been the establishment of local offices of central agencies for the specific purpose of supporting SMEs. To ensure local ownership, these often are partly funded by local authorities or organizations such as the chamber of commerce. The local offices are equipped to provide technical, financial, managerial, and marketing support, drawing as needed upon expertise in universities, public laboratories, or networks of consultants.

Complementing these supporting modalities, incentives have encouraged SMEs to use research and technology facilities (such as public laboratories and university centers) and to stimulate the interactions between research bodies and industry that are at the heart of the innovation process. Most such incentives have been of two types: (a) small subsidies granted to SMEs to reimburse up to half of their contracts with accredited public or university laboratories, or to employ in-house researchers seconded by these laboratories; and (b) large grants for projects developed through university–industry collaboration, on the condition that matching funds are provided by the private sector. The latter grants sometimes have been focused on specific technologies that the government wishes to promote.

Source: Authors.

Table 5.1 *Policy Instruments Suitable for Stimulating Enterprise Innovation*

Segment of the domestic enterprise sector	*Policy objectives*	*Policy instruments*
Low-technology SMEs and microenterprises	**Business:** Stabilize businesses and build competitive capabilities. **Innovation:** Build awareness of scope and benefits of innovation	• Business advisory and support services; SME and microenterprise support agencies • Finance (including microfinance) • Management and skills development • Innovation awareness and understanding • Productivity-enhancing services • Innovation identification and matchmaking • Cluster-based approaches to stimulating innovation
SMEs with modest technology	**Business:** Develop competitiveness. **Innovation:** Introduce basic innovation skills. Encourage adoption and application of new ideas	• Support for business development, diversifying customer base • Product diversification and quality improvement • Management and skills development • Internet-based information services • Technology awareness and marketing • Support for technology adoption and adaptation projects • Graduate internship and placement programs • Consultancy and technical assistance support

Table 5.1 *Continued*

Segment of the domestic enterprise sector	*Policy objectives*	*Policy instruments*
Technologically competent enterprises	**Business:** Support market development, internationalization of businesses. **Innovation:** Build in-house innovation capabilities.	• Business development, export market support • Internet-based information services • Innovation and technology support • Technology transfer support • Incubators and technology parks • Linkages with academic researchers • Innovation relay centers, matchmaking services • Laboratory services and metrology • Graduate internships and placements • Consultancy and technical assistance support for commercialization, intellectual property rights, licensing, patenting, etc. • Technology joint ventures
R&D-rich enterprises	**Business:** Develop international markets, promote entry to global supply chains **Innovation:** Encourage R&D, engagement with international innovation networks, and technology transfer and diffusion	• Export support • Technology support • Support for participation in international R&D networks such as the EU 6th Framework Program • Technology and other innovation-based spin-offs • University-industry collaboration • Support for commercialization

Source: Authors.

Three major kinds of support are needed to spur innovation in developing countries: financial, technical, and regulatory.

Financial support. Firms and local communities often lack the resources needed to accelerate the design, testing, use, and dissemination of technologies. In such cases, two complementary schemes can be useful. Both are based on simple matching-fund principles and provide 50 percent of the funding required for the development phase of small and medium-sized projects (up to US$20,000, for example).

- "User schemes" allow groups and communities to buy needed technologies, providing, if appropriate, complementary in-kind support (such as labor).
- "Developer schemes" fund 50 percent of technical services or R&D projects undertaken by SMEs in partnership with research organizations (such as public laboratories and universities). Schemes of this type enable research organizations to use their knowledge and resources to serve communities indirectly but effectively.

Such support schemes should be managed primarily at the subnational level, with the national authorities providing oversight only. Regional and local commissions can screen and select projects with support from outside experts (including foreign experts) if necessary. Local management avoids the bureaucracy that would burden such schemes if administered at the central level.

Technical support. Based on experience in advanced countries, a network of locally based and owned structures should be set up to serve the needs of rural and urban communities for technical advice, information, and assistance (in design, marketing, etc.). Extension services in agriculture and design and manufacturing workshops for industry can be established, based on demand from local communities and cofunded (for example, on a 50/50 basis) by local organizations (municipalities or associations of firms or farmers) and the government. Such services may be operated as branches of central bodies, with appropriate sharing of data and other information.

Regulatory support. Regulatory action is needed to deal with several issues, including those described below.

- *Transparency in the use of public resources.* Clear legal and administrative procedures should govern actions to stimulate service-based R&D contracts and to formalize linkages between the business sector and the R&D infrastructure. Regulation is needed because such relationships may involve the use of public equipment and personnel by firms (possibly including the temporary employment of public researchers), as well as intellectual property rights.
- *Health, safety, and quality standards.* Developing-country exporters of processed food and other products may have to meet high standards in import markets. Government action can help raise quality awareness, improve quality control, and put in place systems of accreditation and certification that help domestic firms meet international standards.
- *Access to credit.* Credit guarantees and other regulatory mechanisms can help alleviate financial problems for firms and individuals unable to obtain credit through the banking system.

Research and Technology Infrastructure

A first layer of "soft infrastructure" is needed for the support of innovative initiatives. It consists of basic activities such as metrology, standards and quality control, extension services (for manufacturing and agriculture), information and training programs, and demonstration and pilot projects. Crucial to the efficiency of such schemes, as discussed previously, is their proximity to local entrepreneurs, potential innovators, and the population. Such services are vital public goods and should be funded and managed accordingly. In particular, organizations for technical standards and norms should not be privatized.

Soft infrastructure also involves the building of appropriate research structures. The activities pursued within those structures, from basic to more applied, need to be adapted to local needs and capabilities. A major problem in more than a few developing countries is that research bodies are isolated from local communities. Creating links to end that isolation depends primarily on the conditions of financing for research bodies. A good rule of thumb, based on the experience of industrialized countries, is that 50–70 percent of the research body's budget is provided as guaranteed resources (core funding), with the remaining 30–50 percent taking the form of more precarious resources (contracts). Some developing countries, such as India, have been able to transform large sections of their research structures along these lines (box 5.5).

University research plays a crucial role in industrialized countries. This is not yet the case in developing countries, where universities have focused on training and education. However, academic research in developing countries should gradually come to represent both a search for excellence and a service to the surrounding community. An exceptional example from Mali is presented in box 5.6. The Malaria Research and Training Center at Bamako University is active in top scientific networks at the global level while maintaining its links with groups of traditional healers at the local level.

Programs for the Diffusion of Basic Technology

Lives depend on the successful dissemination of basic technologies in poor countries, as in poor areas of more advanced countries. Improving food production and storage, agriculture productivity, water supply and sanitation, and public health, among other goals, requires packages of actions that provide education and technical assistance while closing infrastructure gaps. The challenge is well illustrated by the recent observations of a World Bank team that addressed science and technology capacity building in Rwanda (box 5.7). The team found that great gains could be made by creating, often very inexpensively, the conditions for the proper use of available technologies.

Measures such as the innovation agencies proposed elsewhere in this chapter can partly resolve the problem of diffusing basic technologies, but they are not really adapted to the task. Crash programs of far broader scope are needed to tackle the knowledge deficit of poor areas. Such programs must take into account the complexity of the situation and the different parameters involved (technical and regulatory), keeping in mind that the beneficiary communities generally lack the financial resources needed to pay for needed changes. It is also crucial to take a long-term perspective when building capabilities in villages to ensure

Box 5.5 *India's Council of Scientific Industrial Research: From Self-Sufficient Technological Development to Competitive Market-Driven R&D*

Set up in 1942, India's Council of Scientific Industrial Research (CSIR) was modeled on the United Kingdom's Department of Scientific and Industrial Research. It predated most other specialized R&D institutes in India and had a wide range of functions, ranging from promotion of scientific research to setting up R&D institutions and collecting and disseminating data on research and industry. Following India's independence from Britain in 1947, the CSIR was made an independent society under the prime minister. In the first two decades after independence it focused on building up an extensive R&D infrastructure from metrology to R&D over a wide range of industries, with a strong focus on supporting emerging industry, especially small and medium enterprises.

The global energy shock of the early 1970s coincided with three consecutive years of drought in India. In the pursuit of Indian self-reliance, CSIR concentrated on reverse engineering of product and process technology—primarily in pharmaceuticals and chemicals, glass, and other import-substituting industries—and on adding value to technologies that made use of domestic resources, such as high-ash coal, small-scale cement plants, and medicinal and aromatic plants.

When India changed from an inward-oriented development strategy to a more outward and market-driven one after the country's 1991 economic crisis, the focus of CSIR also changed. With the liberalization of trade and industrial policy, firms began to feel more international competition. CSIR was criticized for being (a) unwieldy, (b) not very effective at transforming scientific results in the laboratory to technologies for industrial production, and (c) focusing too much effort on known processes (and thereby "reinventing the wheel"). The demands of the crisis led to a self-examination and a radical change in CSIR's role, from one that embodied and expressed India's former goal of technological self-reliance to a new vision of CSIR as a business that responded to the needs of a competitive market with world-class industrial R&D. Greater emphasis was placed on output and performance, and on income-producing work relevant to the productive sectors. Each laboratory was considered a corporate subsidiary, and incentives and rewards for meeting targets were introduced. The laboratories earned their operational autonomy by delivering promised output. Efforts have continued to improve effectiveness and efficiency.

Although CSIR is undergoing further restructuring, its impressive results since 1991 reveal the potential impact of a decisive change in the direction and incentive regime of even a very large public research system. Between 1997 and 2002 CSIR reduced its laboratories from 40 to 38 and its manpower from 24,000 to 20,000. At the same time, its output increased noticeably. Technical and scientific publications in internationally recognized journals tracked by the Science Citation Index rose from 1,576 in 1995 to 2,900 in 2005; their average impact factor increased from 1.5 to 2.2. Patent filings in India increased from 264 in 1997/98 to 418 in 2004/05. Patent filings abroad increased from 94 in 1997/98 to 500 in 2004/05. CSIR accounted for 50–60 percent of all U.S. patents granted to resident Indian inventors. Contract income grew from 1.8 billion rupees in 1995/96 to 3.1 billion rupees in 2005/06 (about US$65 million).

Today CSIR has 4,700 active scientists and technologists in 37 research laboratories supported by 8,500 scientific and technical personnel. Since 1997 its core budget has doubled to 15 billion rupees (about US$325 million), so its earnings add about 20 percent to its core budget.

Source: Adapted from Bhojwani 2006.

Box 5.6 *The Malaria Research and Training Center of Bamako University in Mali*

Created in 1992, Bamako University's Malaria Research and Training Center (MRTC) is internationally recognized for its contributions to research on malaria and improvement of public health standards. The MRTC has benefited from a clear strategy with ambitious goals, strict implementation methods and tools, and wide local and international partnerships. It has benefited from a supportive social and political context.

Roots. In 1976, Professor Philippe Ranque of the University of Marseilles started a department of epidemiology and parasitic disease at Bamako University, which he ran with three well-trained and motivated researchers. After Ranque's retirement in 1988, his team decided to scale up the department's research capacity. High demand for work to fight malaria encouraged the Malian government to support the creation of the MRTC. The center was first enlarged to 12 Malian researchers and later to 24 African researchers.

Strategy. MRTC's mission was to increase the understanding of malaria in Mali and elsewhere in Africa through research and training; to produce high-quality scientific results to inform decision making; to strengthen the implementation of public health programs; and to participate in international efforts to develop vaccines and medicine. Its strategy was three-pronged:

The quality requirement. Quality is ensured through selective enrollment and cooperation with universities in advanced countries. Research is funded through international competitive grants (60 percent of MRTC's budget).

The implementation tools. Management is based on recognition of merit, cooperation, and confidence-building among team members. Research and administration are kept separate. Research is conducted under international standards (with Internet access, a satellite connection, and adequate laboratory equipment), thus preventing brain drain and helping to ensure top-quality research. A management style that emphasizes individual and collective responsibility ensures motivation and cohesion.

Partnerships at the local, national, and international levels. MRTC has developed partnerships far and wide, seeking synergy among the actors involved in fighting malaria. It collaborates with local authorities and doctors who know local needs and behavior. Field research is conducted with the cooperation of local people. Government support ensures legitimacy and allows MRTC to develop other partnerships. MRTC's international activities include cooperation with universities, research centers, and international agencies. Overall, the importance of the goals pursued as well as the results obtained have helped MRTC raise funds and boost its research capacities. MRTC has opened its doors to students from other African countries and collaborates with units at other African universities, thus encouraging the diffusion of research excellence through partnership networks and the development of local capacity.

Achievements. In 2006, the center had a faculty of 30 professors, plus 50 Ph.D. students/research assistants and 40 administrative staff. It has gained international recognition, having published more than 200 articles in international scientific journals since 1992. It has succeeded in international grant competitions and been hailed as a center of excellence by the Agence Universitaire de la Francophonie and the U.S. National Institutes of Health. MRTC is also certified by the U.S. Food and Drug Administration to conduct clinical tests according to international standards; tests of anti-malaria vaccines will start in 2006. MRTC researchers have organized with traditional doctors a network for immediate care for infected persons in the Bandiagara region, an initiative that has significantly reduced malaria mortality. In 1997, prior to the program, the mortality rate among children under 5 years of age was 20 to 30 percent. By 2005 it had been reduced to between 5 and 7 percent.

Source: Adapted from Doumbo 2005.

that programs will continue long enough to have the desired effect. This goes in particular for the building of vocational schools to provide a minimal technical culture among populations, and particularly among the young.

Crash programs to close knowledge deficits can be included in larger donor-supported projects in individual sectors. However, as the case of Rwanda illustrates, it is necessary to address simultaneously the needs of poor communities for knowledge and basic technology related to food, water, health, and other basic needs. All these issues need to be tackled at the same time to ensure the survival of the communities.

Export Sector Policies

A recent World Bank study (Chandra 2006) of 10 export sectors in low- and medium-income countries—oil palm and electronics in Malaysia; electronics in Taiwan; maize, grapes, and software in India; fish in Uganda; cut flowers in Kenya; and salmon and wine in Chile—casts light on the principles that should guide government policy toward potential export sectors, identifies the role of knowledge and technology in developing those sectors, and explores the channels through which technology transfer operates.

Some key principles in the development of successful export sectors are: common goal targeting (such as accumulating foreign currency or creating jobs); political vision; rewarding winners and abandoning losers (to establish clear incentives); private-sector-led export development (to ensure that the whole process addresses world market requirements); maintenance of strong internal competition; and strict respect of property rights and the rule of law. The study finds that there is a clear need for government support but warns that the public role must not interfere with market-economy principles.

The industry-specific interventions observed in the cases treated in the study fall into the following categories: negotiating with multinational corporations, spinning off domestic firms, facilitating acquisition and dissemination of technologies, promoting exports, developing industrial clusters, providing regulatory services, supporting industry organizations and coalitions, and meeting technical manpower requirements. Government has a significant role to play by providing legal, organizational, and human support, which is far more crucial than direct financial support.

The cases reveal that foreign technology is transferred through many channels—among them FDI, licensing agreements, and imports of capital goods, depending on the sector—but that success depends in every case on active involvement and participation of local industry (cooperation, training) and on the development of local R&D. There is clearly a need to develop local receptiveness to technology. The study observes that technology transfer from MNCs is "tricky." Hence the importance of programs such as the Irish Linkages Program profiled in box 5.2.

It is sometimes useful to have a specialized agency focusing on the promotion of dynamic, export-oriented industries. The case of Fundación Chile (box 5.8), which was directly involved in the development of that country's wine and salmon industries, is an exceptional success story that may provide guidance for nascent innovation agencies elsewhere, including those of broad scope required in countries where knowledge and technology are less well developed than in Chile.

Box 5.7 Basic Technology and Capacity-Building in Rwanda

- **Agricultural productivity.** The cultivation methods and other technical know-how needed to boost agricultural productivity are widely known around the world—but not by Rwanda's subsistence farmers. The Millennium Village project in Mayange shows that yields can be raised substantially through simple improvements in cultivation practices, the use of improved seed varieties and fertilizers, and easy measures to improve water use and prevent soil erosion. In Mayange, most of these practices were introduced into the village by one well-trained local extension worker. Nearby villages, as well as farmers who initially chose not to accept the new inputs and techniques, are now clamoring to participate. Scaling up the project's success depends on training and motivating extension agents, providing enough packages of the improved inputs, and diffusing known cultivation techniques to additional villages and farms. Many of the extension workers now working in the Ministry of Agriculture are inefficient and poorly utilized. Many simply do not have what it takes to teach local farmers how to enhance agricultural productivity.
- **Alternative energy.** Most Rwandans live in villages that are not connected to the power grid. Building central generating plants and connecting remote villages to the central grid is neither feasible nor affordable. Therefore, Rwanda will need to develop alternative, decentralized energy sources that include wind, solar, thermal, small-scale hydropower, and bio fuels. While every home cannot be connected to these alternative energy supplies, one or two electrified buildings in each village can serve as schools, computer centers, libraries, public health facilities, and agricultural extension centers.
- **Water conservation and rainwater harvesting.** Water-borne diseases caused by a shortage of potable water are a major source of illness in Rwanda. Rainwater harvesting can provide a relatively low-cost and effective water supply for cooking and drinking. The techniques and technologies for harvesting rainwater are widely known but not widely used in Rwanda. Part of the problem is that the local population does not have the technical and vocational skills needed to build systems to collect rainwater. A vocational training program to boost the supply of trained technicians, along with a program to finance the construction of rainwater harvesting systems, would help to address both the supply and demand sides of the equation.
- **Food processing and food storage.** Increasing agricultural yields will not improve food security if food cannot be safely processed and stored. New technologies need to be developed and deployed to process and store food without large amounts of scarce electricity.
- **Public health.** The rural population need education in areas such as nutrition, sanitation, hygiene, disease prevention, and the importance of clean, safe drinking water. Public health extension agents and medical technicians must be trained, deployed in villages, and connected to regional health centers via the Internet. All this will entail targeted capacity building.
- **Technical and vocational education.** Rwanda suffers from a major shortage of technicians who can repair and maintain automobiles, electrical appliances, and electronic equipment (such as printers and copiers), and of craftsmen who can design and build rainwater harvesting systems, schools, and other facilities. Technicians and craftsmen are scarce in part because there is a shortage of well-equipped technical and vocational schools, which cost more to operate than academic secondary schools. Donors seeking to maximize the number of students who benefit from aid-financed education programs often prefer to invest in basic secondary education, with its lower unit costs, even when this does not answer the economy's most urgent needs. Rwanda will have difficulty moving beyond subsistence agriculture without an adequate supply of personnel trained in mid-level technical skills.

Source: Al Watkins, World Bank.

Box 5.8 *The Fundación Chile Model for Promoting Innovation*

Fundación Chile is one of the most successful attempts in the Latin American region to establish national bodies for the promotion of new technologies. Founded as a joint effort between the Chilean government and the U.S. firm ITT, it is now largely autonomous. Fundación Chile uses four main techniques to stimulate the transfer and dissemination of technology. First, it creates innovative enterprises, almost always in association with companies or individuals. Second, it develops, adapts, and sells technologies to clients in the productive and public sectors, both in Chile and abroad. Third, it fosters institutional innovations and new transfer mechanisms. Fourth, it disseminates technologies to multiple users though seminars, specialized magazines, project assistance, and so on.

Fundación Chile focuses on biotechnology, management, environment, and financial engineering and information, among other areas. Recent areas of concentration include forestry genetics and DNA vaccines for aquaculture. Once Fundación Chile identifies a cluster with comparative advantage it focuses on supplying the missing links that will help the cluster take off. Past and present clusters include agribusiness, marine industries, tourism, and forestry and wood processing. The Fundación also applies the missing-link concept when supporting education and human resources development.

Fundación Chile has been quite successful in incubating new ventures through entrepreneurship and technological innovation. Some of the demonstration companies established by Fundación Chile have failed, but overall the practice has proved effective as a method for disseminating new technologies. Companies are transferred to the private sector once the technologies that underpin them have been proven in practice and their economic profitability established. By 1999, Fundación Chile had launched 36 such ventures, 17 of which have been sold. The six leading companies have generated more revenue than the total cost of the Fundación during its existence. One of the most successful cases, which exhibits many elements of the successful development of a knowledge cluster, is that of the salmon industry, which became a dynamic export sector in Chile over a period of ten years.

Among the success factors identified by Fundación Chile are:

- An entrepreneurial, highly paid, and highly professional management team that may take years to establish
- Systematic monitoring of new technologies available worldwide to bring in those that are most relevant to domestic needs
- Arm's-length relationships with the government, so that the innovation agency and the ventures it supports operate as businesses, not as a public sector organization
- Private shareholders who do not expect an immediate return and tolerate risks.

Source: Authors.

Intellectual Property Rights

Worldwide, the capability for the development of science and technology is very asymmetrically distributed. R&D spending has increased steadily in the industrialized countries since the 1970s, with growing private sector participation. In many of these countries, private firms account for half or more of R&D spending. Developing countries, on the other hand, account for only about 4 percent of global R&D expenditures. The asymmetry is also illustrated by patent registration statistics: industrialized countries hold 97 percent of the world's patents (UNDP 1999).

The international framework for IPRs affects the conditions for access to and use of technology and, therefore, the patterns of industrial and technological development in developing countries. The last decade saw an unprecedented increase in the scope and level of protection of IPRs. New rights are being created, and standards are being harmonized throughout the world. In terms of collective action, the 1994 Agreement on Trade-Related Aspects of Intellectual Property Rights (TRIPS), reached during the Uruguay Round of international trade talks, extended and enlarged the minimum standards that countries must apply to protect IPRs. Continuing discussions in the World Intellectual Property Organization (WIPO) aim at further harmonization of national patent systems and may supersede the TRIPS provisions. Meanwhile, bilateral and regional trade and investment agreements between industrialized and developing countries often include mutual commitments to implement intellectual-property regimes that go substantially beyond the minimum standards of the TRIPS agreement.

From all quarters, therefore, the pressure is growing on developing countries to increase levels of IPR protection in their own regimes based on the standards applied in industrialized countries.

Developing countries that wish to attract technology-rich investment do need to establish effective systems for enforcing IPRs. In many cases (software is an example) the scale of estimated losses from illicit copying is far higher in industrialized countries than in the developing world. Even so, weak levels of enforcement have undoubtedly had a strong impact on the diffusion of knowledge and knowledge-based products in some areas, with the perverse result that many poor people in developing countries have been able to access protected software and other works only as unauthorized copies (usually available at a fraction of the price of the authorized editions). An inevitable impact of stronger copyright protection and enforcement, as required by TRIPS, may be to further reduce access to knowledge-related products in developing countries, notably educational materials, with potentially damaging consequences for the poor.[2] Open source software may provide a partial solution for developing countries (box 5.9).

Discussions of the appropriate role of IPR policy for development tend to be clouded by the heterogeneous nature of the technical and scientific capacities of developing countries (UNCTAD and ICTSD 2003). The more technologically advanced among them may wish to adopt systems that provide extensive patent protection as incentives for R&D. However, they should avoid those aspects of the system that might provide disincentives for R&D or lead to the diversion of resources to litigation and disputes about patents of doubtful validity. Such systems must have adequate safeguards to ensure a competitive environment and minimize costs to consumers. This is especially important in areas such as pharmaceuticals and agriculture, where the cost of providing strong patent protection is likely to be greatest (Barton 2003).

For the majority of developing countries, especially low-income countries, the best system may be one that applies strict standards of patentability and results in fewer inventions that meet the relevant criteria. A second tier of protection can be found in so-called utility models, which offer protection under less stringent criteria. Such protection may be more appropriate for the economic circumstances of many developing countries (CIPR 2002; Aubert 2005a). Moreover, some developing countries have benefited from copyright protection. The software, music, and film industries in Brazil, India, Jamaica, Mexico, and Thailand are good examples.

---

2. The cost of software is a major problem for developing countries and the reason for the high level of illicit copying. Copyright can also be a barrier to the further development of software to meet local needs and requirements.

Box 5.9 Open Source Software: An IPR Regime Friendly to Developing Economies

The growing commercial and creative success of open source software challenges copyright as the key enabler of sustainable innovation. It focuses on the right of access, not the right to exclude or protect. The common denominator of open source licenses is freely shared source code. Anyone can contribute to an open source project, customize the software to meet new requirements, or take its development in a different direction—and the results of such modifications remain open to others. Open source business models nurture collaborative innovation to create a freely accessible knowledge base in hopes of making a profit on other services, usually with higher value added. For example, some companies profit on installation, customization, or customer support. Others offer packages of open source and proprietary software or hardware.

The open source model and the formal system of intellectual property rights (IPRs) are at opposite ends of the spectrum. The IPR system rewards original innovators and content creators, whereas open source empowers users and promotes follow-up innovation. Striking the right balance between the two approaches will be complicated and messy, but ultimately it may create a more dynamic innovation process, one that is more equitable and less restrictive.

Countries accustomed to tolerating piracy should keep in mind that open source does not mean the end of enforcement of IPRs. Quite the contrary: if businesses ignore the license agreements and use open source software to create proprietary products with closed codes, they undermine the very fundamentals of a sustainable open source movement.

Open source products have a number of advantages. The cost advantage can be very significant. In terms of software, for example, the prohibitively high cost of proprietary products for an average user in developing countries often makes open source software the only realistic alternative to piracy. Unlike proprietary offerings, open source products can be modified, which makes them superior in terms of customization and localization for developing countries. By allowing reverse engineering and collaborative product development with peers from around the globe, open source software can play a very important role in the education and capacity building of specialists in emerging markets.

Source: Authors.

On the periphery of IPRs as understood in the industrialized world, developing countries have a number of specific concerns. In the key area of access to essential medicines, for example, the 2001 Doha Declaration on TRIPS and Public Health recognizes certain conditions under which IPRs may be relaxed or legally contravened. Another IPR-related area important to many developing countries is the protection, preservation, and promotion of indigenous knowledge, which is increasingly recognized as a valuable asset in industrialized and developing countries alike.[3]

3. There are a number of reasons for protecting and promoting traditional knowledge. They include the erosion of traditional lifestyles and cultures through external pressures, misappropriation, and the preservation of biodiversity and promotion of its use for development purposes. Some seek to protect traditional knowledge against commercial exploitation; others to ensure that it is exploited equitably for the benefit of its holders. Underlying the debate on the protection of traditional knowledge may be broader issues such as the position of indigenous communities in the overall economy and society of the countries in which they reside, and their access to, or ownership of, land they have traditionally inhabited. The Intergovernmental Committee on Intellectual Property and Genetic Resources, Traditional Knowledge, and Folklore is dealing with these complex issues at WIPO.

Indigenous knowledge provides input in many modern industries—among them pharmaceuticals, cosmetics, agriculture, food additives, industrial enzymes, biopesticides, and personal care. Yet most of the value added in such cases is appropriated by firms in industrialized countries, whose advanced scientific and technological capabilities make appropriation possible without the prior informed consent of the holders of that knowledge (CIPR 2002).

The core implication of the analysis of IPR-related issues is that the interests of developing countries are best served by tailoring their IPR regimes to their particular economic and social circumstances. A crucial question, however, is how to accommodate this within the current international architecture of the multilateral, regional, and bilateral rules and standards that limit the freedom of countries to act as they choose. With the rapid development of international rules on IPRs,[4] it is important that their actual and potential impact be properly understood if policy making is to be more firmly based on evidence and less on preconceptions about the good or harm that such rules may imply for developing countries.

4. A year or so after TRIPS was agreed, WIPO completed two new international treaties concerning copyright and the Internet.

6

Strengthening Information and Communication Infrastructure

A strong information and communication technology (ICT) infrastructure enables rapid and efficient exchange of information around the world. It also offers the promise of addressing the fundamental knowledge gaps that characterize the low growth rates and persistent poverty found in many countries. Recent advances in various technologies are affecting the acquisition, creation, sharing, and application of knowledge, which in turn affect a wide range of economic and social activities—among them how manufacturers, service providers, and governments are organized and how they perform their functions. Applications of ICT are improving services and creating new opportunities. In particular, greater efficiency and the innovation inherent in new technologies are changing manufacturing, trade, governance, health care, agriculture, and the delivery of government services. As knowledge becomes an increasingly important element of competitiveness, ICTs are reducing transaction costs, eroding time and space barriers, enabling the mass production of customized goods and services, and substituting for limited factors of production.

While a strong ICT infrastructure can be a powerful enabler of productivity, development, and growth, it is important to maintain a realistic understanding of its benefits and limitations in an overall knowledge economy (KE). ICT is not a silver bullet that will solve the problems of low- and middle-income countries and guarantee their success as knowledge-based economies. Gone are the days when ICT was considered a viable means of leapfrogging from one level of development to another. It has become essential to consider a number of other factors that affect how economic growth is created and sustained—chief among them an enabling environment that encourages market competition, innovation, and private sector growth and productivity. *ICT can help transform the economy, but it is no replacement for the larger changes often necessary at the policy and institutional levels.* In short, a dynamic and efficient ICT infrastructure can enable change, but will not create change by itself.

The prerequisites for benefiting fully from ICT can be seen in figure 6.1, which illustrates the deep but fragile connection between the value derived by ICT users and the broader enabling environment. This environment includes legal and regulatory frameworks, access to capital, conditions for and constraints on new business creation, the strength and flexibility of domestic financial institutions and markets, human capital, and so on.

In simple terms, the potential value of ICT is what people actually do with it. Use requires access, and, to provide access, a country must become "e-ready." Clearly, there can be no access without an adequate supporting infrastructure.

Figure 6.1 Prerequisites for Capturing ICT-Induced Benefits

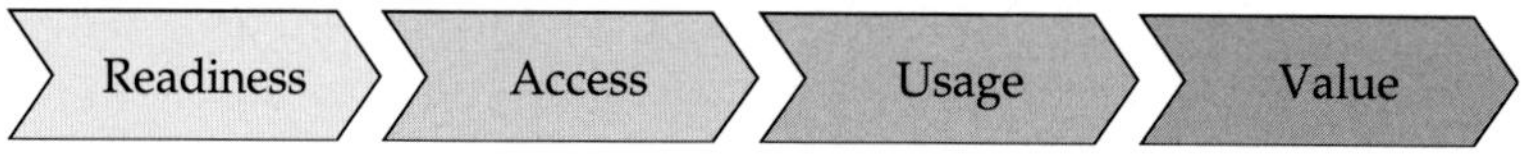

Source: World Bank 2006e.

To create and sustain a strong overarching ICT infrastructure, it is imperative to mobilize the many stakeholders involved in its deployment and use: government, businesses, universities, telecommunications and information providers, rural representatives, and so on. Without the genuine buy-in of these diverse and influential groups, strategies and plans will fail to reflect the needs and challenges of the country, and implementation will likely falter as a result.

A Holistic Approach

The key elements of ICT-enabled development and growth are highly interdependent and complemented by macroeconomic, financial, educational, and other policies. Governments tend to establish independent, sector-based ICT projects, however, which may undermine broader strategic goals by diverting attention and resources away from them. *Building a strong ICT infrastructure using a holistic approach is unequivocally the key to success.* Such an approach takes into account the impact of interventions at the macro level, facilitates interactions across sectors and agencies, promotes realistic implementation strategies, and is relevant to local conditions. It also allows for adaptation, learning, and innovation over time.[1]

A number of countries have embarked on "e-strategies" that serve as holistic plans of action for how ICT should be developed and used to achieve broader development objectives. Although not all have been successful, several, such as those of Estonia, Mozambique, Rwanda, and Sri Lanka (box 6.1), are demonstrating the benefits that can be attained through such an approach.

A holistic approach also strongly embraces the entire range of available tools to ensure greater impact. The past 10 years have witnessed an explosion of mobile telephone and Internet use that is having a fundamental and lasting impact on how people conduct business, engage in lifelong learning, and interact with each other and with their governments. However, these technologies are not the only ICT tools people have at their disposal. After all, *old* does not necessarily mean obsolete. In fact, a quick way of lowering the cost of technology is to rely on older and oftentimes more proven means of delivering information. Radio, television, and print media continue to be important and reliable sources of receiving and sharing information. Radio broadcasting covers 95 percent of the world's population, and television 85 percent. Community radio is an extremely powerful tool that reaches thousands of small and poor communities unfamiliar with other forms of ICT. Also, in many villages around the world, few people sit in front of a computer. Instead, they receive much of their information through loudspeakers from village leaders or from newspapers.[2] Examples of the impact of these older but proven technologies are abundant (box 6.2).

---

1. Much has been written on the need for a holistic approach. See chapter 1 of *E-Development: From Excitement to Effectiveness* (World Bank 2005b) for a comprehensive look.
2. The use of a wide range of ICT technology was covered in *The Economist* (March 10, 2005).

Box 6.1 "e–Sri Lanka" and a Holistic Approach to ICT

Sharing a strong belief that ICT enables development and global competitiveness, Sri Lanka's government, private sector, and other stakeholders have developed a vision for "e–Sri Lanka." This involves introducing ICT to every village, citizen, and business, and revolutionizing the way the government operates. A concerted effort is being made to design and implement a comprehensive, nationwide strategy to harness the potential of ICT to achieve broader socioeconomic goals. In collaboration with international donors, the project attempts to fulfill these goals through affordable access to ICT, modernized government practices, delivery of citizen services, the development of public–private partnerships, and the creation of competitive knowledge industries. Adopting a comprehensive and inclusive approach, the program cuts across many sectors and has devoted much attention to developing the skills of government workers, students and teachers, and other citizens.

Source: Development Gateway; e–Sri Lanka Web site.

Box 6.2 Using "Older" ICT in Three Countries

Community Radio (Ecuador)
Radio Chaguarurco was launched in 1996. It is located in the province of Azuay, a mountainous region mainly populated by farmers. As a self-financed station that makes heavy use of volunteers, Radio Chaguarurco produces programs that explain complex issues in everyday language. The themes covered include health, environmental, cultural, and human rights issues. A weekly program that examines agricultural and consumer processes and trends has played an important role in controlling prices and speculation. Evaluation results have been very positive. One farmer was quoted as saying, "The radio is the telephone for those who don't have one."

Television (India)
The explicit use of television to impart literacy skills through song is unprecedented in India, and is relevant wherever music videos are shown on TV and literacy skills are low. The program on "Same Language Subtitling on TV for Mass Literacy" builds on people's existing knowledge of lyrics, enabling early literates to anticipate the lyrics and read along; the inherent repetition in songs makes them an ideal vehicle for practice. The use of subtitling is a simple approach that leverages popular culture to encourage many Indians to read. It also helps the deaf and hard-of-hearing by making television programming more accessible. The possibility of replicating this inexpensive approach in India with different languages and in other countries is enormous.

Walkie-Talkies (Uganda)
The Rural Extended Services and Care for Ultimate Emergency Relief program was launched in March 1996 in the Iganga District of Eastern Uganda. Initiated by the Ministry of Health, United Nations Population Fund, and the Uganda Population Secretariat, it was designed to link traditional rural community health providers with the formal health system in a cost-effective way. Traditional health providers were given walkie-talkies as a way of contacting nurses and physicians if they encountered complications in a delivery. During the initial three-year period, the increased number of deliveries under trained personnel and increased referrals to health units led to a reduction of about 50 percent in the maternal mortality rate in the district.

Source: Girard 2003; Development Marketplace; The Communication Initiative.

Also, scaling up other forms of proven technology such as voicemail, personal digital assistants (PDAs), voice over internet protocol (VoIP), and wireless fidelity (WiFi) could contribute immensely toward addressing access gaps in many areas and provide the means for basic and new services. The potential of WiFi in expanding access to nearly every region in a country, no matter how remote, is nearly limitless, and the cost of deploying it can be extremely low when done in the context of a community-based network. Open source software, with its inherent adaptability to local needs, innovative qualities, and low costs, also offers considerable opportunities to low- and middle-income countries.

Key ICT Dimensions

It is very difficult to generalize about any aspect of developing countries since they vary enormously in size, population, resources, culture, and the nature and extent of the challenges they face. Several poor but rapidly emerging countries, such as India and China, are well on their way toward creating a strong ICT infrastructure that fuels economic growth and delivers significant public services despite widespread poverty in many areas. Other low-income countries, such as Senegal and Mauritania, also score relatively high in this area, even though they rank well below the rest of Sub-Saharan Africa in other measures of development. Some middle-income countries face substantial challenges while others, including Brazil and Jordan, are successfully expanding the range of their ICT-enabled activities. What can be generalized is the need to "get the basics right." This message applies to all developing countries and entails, first and foremost, establishing the underlying conditions for success. Based on the experiences of countries that have made significant progress in this area, these conditions usually involve a strong legal and regulatory framework, the active role of the private sector in the ICT agenda, effective leadership from the government, strong public support, the commitment of political elites, effective pilot programs that build further support, and the strong institutional capacity and coordination of government agencies working on ICT.

Strengthening ICT infrastructure, an integral part of a KE strategy, inevitably requires a critical examination of a core set of dimensions, including:

Enabling environment. Policy, legal, market, and social factors interact at the domestic and global levels to create fertile conditions for ICT-led growth. The enabling environment encompasses national strategies, plans, and leadership, as well as institutions and regulatory frameworks that establish the necessary conditions for access and usage.

Access. Specific results (programs, partnerships, and so on) emerge from the enabling environment, allowing for greater access to ICT, particularly in rural areas, where 70 percent of the people in developing countries live.

Usage. Once access is opened up, services, applications, and content become available for use.

ICT skills development. Human capital is needed to build and maintain a strong information infrastructure.

The objective of the analysis below is to elaborate on these key dimensions of the ICT landscape across low- and middle-income countries, focusing on challenges and opportunities.

An Enabling Environment

Many low- and middle-income countries should establish the requisite foundations of an enabling environment that reduces the obstacles to realizing the potential of a strong ICT infrastructure. In particular, this involves the following areas:

Overall investment climate. Political stability, involving respect for the rule of law and clear and transparent governance structures, is undoubtedly helpful in facilitating the infusion of foreign direct investment (FDI), as is a legal framework that protects investors and consumers. Since 1988, more than 80 developing countries have privatized their telecommunications providers, thereby raising the FDI needed to finance their ICT plans. FDI also plays an important role in sharing knowledge—low-income countries have often taken advantage of the knowledge introduced by more developed countries.

Legal and regulatory framework. If there can be no access without infrastructure, it is also true that there can be no investment without an autonomous regulatory agency that ensures fair competition over the long term, eliminates unnecessary obstacles to market entry, enforces rules and contracts, offers guarantees and dispute-resolution mechanisms to investors, and protects consumers. In fact, telecommunications reform is effective only if the government creates an independent regulator able to support ICT penetration and expedite the deployment of the most appropriate and cost-effective technologies with minimal interference. Creating such an environment is certainly not an easy process. For example, out of 47 countries in Sub-Saharan Africa, only 10 lacked a separate regulatory agency in March 2003. In reality, however, the majority of these agencies possess little independence, weak capacity, and ill-defined mandates largely owing to their strong ties to incumbent telecommunications operators.

National agendas and leadership. It is crucial to have a national policy that supports the development of ICT in the context of broader development objectives, using a holistic approach, and that takes into account the need to facilitate investment, ensure competition, expand access, and identify new areas of need. Leadership at the top can be instrumental in formulating and implementing policy and, in fact, it is difficult to find an example of successful ICT policies in countries where national leaders were not early visionaries and champions or were not actively engaged from the start. Rwanda (box 6.3) is an interesting example of both: a president who had championed the role of ICT developed a consensus among different groups, especially political elites, and pushed forward a strong vision and specific policies.

Institutional roles and capacity. Many countries have failed to deliver on their ICT strategies, often because the absence or weakness of institutional structures made effective implementation very problematic. Even when institutions do

Box 6.3 *Rwanda: ICT Vision and Leadership*

Rwanda is early in its ambitious Vision 2020 program, approved by the cabinet in early 2000. At the policy level, ICTs have been anchored securely into broader economic, social, and development policies and strategies in the form of the National Information and Communications Infrastructure (NICI) plan. Policy and planning are mirrored by a comprehensive process of institution creation and development. The National Information Technology Commission (NITC), a set of working groups, and the Rwanda Information Technology Authority (RITA) are the main coordination entities. The latter, a well-respected organization with a broad range of responsibilities, implements many NICI tasks under the current five-year plan. A single regulatory body, the Rwandan Utility Regulatory Authority (RURA), set up in 2003, is responsible for establishing a Universal Access Fund.

NITC is a high-powered ICT policy think-tank whose mission is to lead the process of creating the Rwandan information society and economy in line with the aspirations of the Vision for Rwanda. It is responsible for advising the government on how Rwanda can best formulate, develop, and implement its ICT policies, strategies, and plans to accelerate the process of transforming Rwanda into an information-rich, knowledge-based society and economy. The NITC is chaired by President Paul Kagame, who is well-known as the national ICT champion.

Source: O'Siochru 2006.

exist, there is often no clear division of responsibility, and this can create political and bureaucratic obstacles and undermine policy coordination and resource allocation. There is no perfect model for an ICT institutional framework, and every country will require a different one. Some of the more successful countries, however, have established an umbrella agency that takes into account the cross-cutting nature of ICT, coordinates the work of other agencies and ministries while mainstreaming the ICT agenda, establishes the coherence of policies across sectors, and adapts to new realities. For some countries, the organization of a consultative body or ICT agency that develops and implements an ICT strategy and policies—and reports directly to a prime minister or president—has been effective.

Important policy changes have taken place in several countries, reflecting improvements in the enabling environment. The result has often been liberalization and increased competition in the telecommunications sector. One example is Mauritania (box 6.4), which undertook major reforms that enhanced the capacity of the country to implement a new telecommunications reform policy.

Other positive developments have occurred. About one-half of all low-income countries have opened their telecommunications markets to competition, leading to growing markets, lower costs, greater innovation, and customized services for different groups of users. Some 130 of 164 countries have at least three competing providers of mobile services. Ten years ago one million phones were available in all of Africa; now there are well over 100 million, mainly mobile. In addition, Internet use has also grown rapidly; the number of users increased by more than four-fold between 2000 and 2005. Despite signs of considerable progress in many developing countries, however, more needs to be done to ameliorate the enabling environment. Greater efforts can be made to liberalize the telecommunications sector, dismantle

Box 6.4 Overcoming Capacity Constraints for Telecommunications Reform in Mauritania

As a result of Mauritania's 1998–2001 telecommunications reform, this desert nation of over two million largely nomadic inhabitants overcame capacity constraints to attain unanticipated outcomes. New private investment in telecommunications amounting to $100 million was attracted over two years, equivalent to 10 percent of GDP; telephone line access multiplied 20-fold; 6,000 new telecommunications-related jobs were created in the informal sector in the capital, Noukachott, alone; and a multisector regulatory agency was established that is now regarded as a model in Africa. At the outset, Mauritania lacked critical skills, but now it has become an example for neighboring countries on how to competitively tender utility licenses, effectively regulate utilities in a competitive setting, and privatize a telecommunications operator. Support for capacity enhancement came from relatively modest external assistance, at an estimated cost of slightly over $1 million. Several factors contributed to the country's success:

- Mauritanian policy makers were aware that capacity needs extended beyond telecommunications policy and regulation and utility privatization to the promotion of investment and the strategic use of media.
- With the World Bank as a key partner, parallel awareness-raising and training processes transformed the early vision of a small number of telecommunications reforms into a framework for broad utility sector reforms, with an interministerial committee, a project management unit, and a multisector regulatory agency all playing valuable roles.
- A simple yet highly effective project management tool mapped a detailed 36-month reform package and implementation plan.
- With early ownership of the capacity enhancement process, Mauritanians avoided equating national capacity with technical assistance. Outside consultants were mainly used to fill short-term skill gaps, providing temporary assistance in legal issues, auditing, and technical training.
- Rapidly translating learning into action, Mauritanians were able to effectively adopt basic telecommunications legislation for best practices and separate and corporatize telecommunications entities in record time.
- Mauritanians took responsibility for difficult and innovative decisions as their capacities for evidence-based policy design grew along with their confidence in rules-based decision making.

Source: World Bank 2003e.

state monopolies, promote private investment in infrastructure and services, and increase competition.

Since they typically enjoy greater political stability and business opportunities than low-income countries, many middle-income countries attract greater foreign investment, which has proven instrumental in improving and expanding what is often an already mature infrastructure. The size and potential of such countries' telecommunications markets are also appealing to investors. On the regulation side, Morocco demonstrates how an effective regulatory framework can attract investors and improve infrastructure (box 6.5).[3]

3. A more detailed analysis of the Moroccan case can be found in Wellenius, Rosotto, and Lewin (2004).

Box 6.5 *The Impact of Effective Regulation in Morocco*

The auction of Morocco's second Global System for Mobile Communications (GSM) in 1999 led to significant growth of its mobile sector. This process was strongly supported by the National Agency of Telecommunications Regulation, which has become a regional model because of the transparent and well-planned way in which it awards GSM licenses. Five years after the auction the number of mobile subscribers in the country grew from 150,000 to over eight million, changing Morocco's status from the country with the lowest teledensity in the region to the highest. The confidence of foreign investors was also strengthened, and many of Morocco's neighbors learned from its experience and followed with their own competitive GSM contracts.

Source: Wellenius, Rosotto, and Lewin 2004.

In some cases, the government can play a very active role by launching and sponsoring initiatives that stimulate a wide range of ICT activities, supporting ICT goals within a broader development perspective. The challenge is not so much to identify basic priorities and action as to go to the next level—scaling up programs, creating greater efficiencies, and, most importantly, formulating and implementing polices that use ICT to increase competitiveness in the global economy. For example, the Republic of Korea's information promotion committee was established in the mid-1990s as a special vehicle to overcome budgetary restrictions and promote the country's ICT-related public works. Managed by the Ministry of Information and Communication, and based on government and private sector contributions, the committee created a system for reallocating profits from ICT fields to the ICT sector.[4] The committee's goals have been to promote e-government, roll out a broadband network, support R&D and standardization, and educate human resources in ICT—all important factors contributing to Korea's rise as a knowledge economy.

Middle-income countries, in spite of their greater resources and opportunities, still face major challenges in achieving an environment conducive to the strengthening of their ICT infrastructure. They need to continue and deepen reforms by further liberalizing their telecommunications market and ensuring greater competition in ICT sectors. They also must strengthen the institutions tasked with carrying out ICT policy, raise leaders' capacity in this area, and fill in existing gaps in their legal and regulatory frameworks. The focus here should be on facilitating the growth of electronic transactions as an integral part of business and government functions, improving efficiency, and addressing the issues of security, privacy, technical standards, liabilities, and so on.

A final dimension of the enabling environment involves the need to monitor and evaluate ICT initiatives on a continuous basis. Without a framework of monitoring and evaluation (M&E), it is impossible to assess the impact of ICT and report tangible results. A recent study of national e-strategies indicates that, regardless of income and region, most countries perform poorly in their use of M&E. Very few strategies include specific M&E components, such as institutional and budgetary support. Considering the substantial resources devoted by many developing coun-

4. For further information on Korea's information promotion committee, go to: http://www.ipc.go.kr/ipceng/project/eproject_intro.jsp.

tries to ICT, it is difficult to understand why this is the case. Lessons and experience derived from ICT activities in numerous developing countries strongly suggest that receiving regular feedback helps national leaders design and implement more effective and relevant ICT policies. Experience also shows that an effective M&E system should directly link ICT strategy to development goals at the national level, rather than simply evaluating individual projects. It should also measure results with appropriate tools and incorporate indicators into national ICT plans; this may involve, for example, examining policy goals, strategic priorities, key initiatives, and actions across the corresponding indicators of impact, outcomes, outputs, and deliverables. Finally, the M&E system should produce and disseminate results that can be easily understood and discussed by a variety of stakeholders in the government, civil society, private sector, and donor organizations.[5]

Access

Access to telecommunications infrastructure for all households is obviously not a realistic goal for most low-income countries for some years to come. But improving public access—particularly for a significant percentage of the poor and isolated—is within the reach of most. In fact, nearly all low-income countries have launched programs to support such an effort. Some examples include setting up universal access funds that encourage infrastructure development in rural areas, and establishing community telecenters and other public access points that offer ICT access and services to underserved areas. Low-income countries have also devised innovative measures to support private and public sector financing and investment, and boosted overall ICT penetration by strengthening partnerships among government agencies, research and academic institutions, nongovernmental organizations (NGOs), and the private sector.

Low-income countries must also address the challenge of making available the hardware needed to improve access. Partnerships, mainly with the private sector, can play an instrumental role here. Ethiopia, as part of its Woredanet project, is working closely with a Belgian company that supplies low-cost personal computers. On a much larger scale, the One Laptop per Child initiative, created by the Massachusetts Institute of Technology (MIT), has the goal of eventually producing 500 million laptops for under $100 apiece. Motorola has also launched an initiative to supply six million phone handsets for less than $40 each, and other mobile phone makers are promising to sell similar handsets in the $20 range by 2007.

While access to ICT is growing rapidly and dynamically, many areas, especially rural and poor urban, remain underserved or not served at all. Estimates from the International Telecommunications Union (ITU) indicate that 800,000 villages, representing 30 percent of all villages worldwide, lack even basic telephone services. In many countries, less than 10 percent of rural areas have access to basic telephone services. Low-income countries in particular should select areas and groups to prioritize as they work to provide full public access—the possibility that everyone, even in a distant village, can get to a phone. Even this modest goal can be problematic and invariably involves (a) the political will and capacity to carefully design and implement policies and specific programs to expand access; (b) the use of cre-

5. A lengthy treatment of M&E can be found in World Bank 2005d.

Table 6.1 Telephone and Internet Access in Low- and Middle-Income Countries
Per 1,000 people

	1980	1990	2000	2005
Mobile telephones	–	0.09	46	258
Fixed telephones	14	27	83	135
Total telephones	14	27	129	393
Internet users	–	–	15	67

Source: World Bank 2006f.

ative, flexible, and gradually phased approaches that facilitate and encourage partnerships, investment, consultation with affected communities, and the integration of new and appropriate technologies such as WiFi; and (c) a willingness to learn from the successes and failures of others, since the collective experience of efforts to expand access is becoming increasingly codified and disseminated.[6]

Mobile phones have enormous potential for expanding access. Some experts have argued that as mobile phones outstrip fixed phones in most developing countries (table 6.1), policy makers should consider achieving universal mobile phone access. One of the first steps many low- and middle-income countries can take to increase mobile phone penetration is to reduce or cut taxes on mobile handsets or connection fees (though this may be difficult for political reasons). As a consequence of its 40 percent tax on ICT equipment (exacerbated by the existence of a monopoly telecommunications operator), Ethiopia has one of the slowest growth rates of mobile phone subscribers in Africa. New mobile subscribers in Bangladesh fell from 11 to 7 percent after a $14 connection tax was imposed, while India saw its overall penetration increase from 1 percent to more than 5 percent over a span of three years following its reduction of handset import duties.[7]

Providing access remains a challenge for most middle-income countries, especially in rural areas and among the very poor. Such countries must continue to make this issue a critical priority and move forward to empower—through ICT access—groups of people who still remain largely excluded from the social, economic, and political fabric of their society. Owing to more extensive experience and the longer time they have had to implement programs, however, such countries' infrastructure agendas are driven less by the need to provide public access for all than by the desire to make access a reality at the household level. Owing to enabling policies implemented by the government, mobile phone use tends to be extremely prevalent, and Internet access—often broadband based—is increasingly available to the wealthy and middle-class, students, and business professionals.

The government is also capable of forging effective agreements to create new programs to expand access. A number of countries, including many low- and middle-income, have established universal service funds (also known as universal access funds) that promote public access to telecommunications services, which are now increasingly oriented toward mobile phones. The funds typically involve subsidizing operators in rural and other underserved areas. The Chilean telecommunications development fund (box 6.6) is one example.

---

6. An excellent analysis of extending ICT access to rural areas and the poor can be found in UNDP (2005).
7. "Making the Connection," *The Economist*, September 29, 2005.

Box 6.6 The Chilean Telecommunications Development Fund

A success story that has inspired many in developing countries is the telecommunications development fund established in Chile in 1994. Its objective was to provide public payphones in rural and low-income urban areas. It offered support to private companies on a reverse or "smart subsidy" basis: contracts were awarded to the operators asking for the lowest subsidy to meet the predefined service requirements.

As mentioned in a comprehensive study, "For every dollar of subsidy, the companies invested up to $20 of their own money, averaging over $6. About $52 million, or one-third of the total investment in rural facilities, was used to install payphones, for which the fund provided subsidies of $22 million. In addition, the companies invested about $109 million to provide other services, mainly individual business and residential telephone connections and value-added services (including electronic mail and Internet access in some areas) for which no subsidies were received."

The fund has been very successful. "Between 1995 and 2000 it supported the provision of payphone service to more than 6,000 rural localities with about 2.2 million inhabitants, thereby reducing the proportion of Chile's population living in places without access to basic voice communication from 15 percent in 1994 to 1 percent in 2002." The government later redefined the fund to support telecenter projects, and the smart subsidy approach has been replicated in many Latin American countries, such as Colombia, Peru, and Guatemala.

Source: Wellenius 2002.

Usage

Access to ICT, while an important factor, does not automatically lead to economic opportunity and social inclusion. Of greater significance than the percentage of the population with access to ICTs is how these technologies are employed and adapted to specific needs. ICTs must be made relevant to user needs and resonate with the people using them. They must improve people's lives, help them to obtain basic government services, and enable them to communicate with family, friends, coworkers, and so on. The failure of many community telecenter projects in poor countries—which have often been better at offering access to services rather than customizing those services for specific community needs—underlines the importance of how people use ICT and what they gain from that use.

Accordingly, emphasis must be placed on understanding and prioritizing the needs of users and potential users, and on how to meet those needs through ICT. While the governments of many low-income countries are well aware of the needs, they can benefit from the input of universities, research institutes, donor organizations, and others in solving the second half of the puzzle. NGOs in particular, owing to their strong presence in many pilot projects, often have a unique perspective on how ICT can be used most effectively. The private sector, with its specific needs, also has a crucial role to play. In Kenya, for example, small and medium size enterprises are playing a leadership role in shaping ICT services for the country as a whole.

Recognizing needs makes it possible to create more appropriate content and services, adapt them to local contexts and languages that add value for their users, and make use of appropriate and available tools. Some approaches successfully use a multitiered approach to introduce the social and economic benefits of ICT to rural populations, taking into account the different capabilities found in

Box 6.7 *Using ICT to Improve the Delivery of Services to Rural Populations in India: Three Examples*

Bhoomi. The department of revenue in Karnataka has computerized 20 million records of land ownership for 6.7 million registrants. Farmers previously had to seek out the village accountant—with accompanying delays, harassment, and bribes to be paid—to get a copy of the record of rights, tenancy, and crops—a document needed for tasks such as obtaining bank loans. Today, for a modest fee of Rs.15, a printed copy of this document can be obtained online at computerized land-record kiosks (Bhoomi centers) in 140 *taluk* offices. Next, all the *taluk* databases will be uploaded to a Web-enabled central database. The record of rights, tenancy, and crops will then be available online at Internet kiosks even in rural areas.

e-Choupal. Launched in June 2000 by ITC, a private company that uses electronic kiosks to source agrocommodities, e-Choupal has become the largest Internet-based initiative in rural India. Its services reach more than 3.5 million farmers harvesting a range of crops, including soybean, coffee, wheat, rice, pulses, and shrimp, in 31,000 villages through 5,200 kiosks across six states. Vernacular Web sites covering each agricultural crop that ITC deals in provide farmers with real-time information on the prevailing Indian and international prices and price trends for their crops, expert knowledge on best farming practices, and micro-level weather forecasts. The e-Choupal model and movement have helped increase aggregate demand by creating a virtual producers' co-operative, thus facilitating access to higher quality farm inputs at lower costs for farmers.

Computer-Based Functional Literacy Program (CBFL). Developed by Tata Consultancy Services, the CBFL works to raise literacy using a mix of methods, such as teaching software, multimedia presentations, and printed material. The method is implemented using computers, which deliver the lessons in multimedia format to learners. Supplementing computers in this process are reference textbooks of the National Literacy Mission. Today, the CBFL project is operational in more than 1,000 centers in Andhra Pradesh, Tamil Nadu, and Madhya Pradesh.

Source: Dahlman and Utz 2005.

poor communities. For example, Internet-based "telemedicine" can be adopted at the district hospital level while simpler technologies such as radio can serve rural health units. Other approaches are more sector driven and attempt to address areas such as education, health, and government efficiency. For example, in 2003, Guatemala piloted a program enabling taxpayers to file their tax return online. The tax agency receives a growing percentage of its revenue from this source. Ethiopia, one of the poorest countries in the world, is spending 10 percent of its GDP on ICT with the aim of linking all of the country's schools and local government offices as part of its national poverty reduction plan. India offers still other examples of how public service delivery has been greatly enhanced for rural populations (box 6.7).

Mobile phones especially are having a transformative effect. A recent report on the results of surveys of rural communities suggests that mobile telephony is accessed by the poorest people, thanks in part to widespread sharing. Individuals surveyed highlighted savings in travel time and costs, easier communication with family and friends, and access to business and job information. A majority of small businesses reported increased sales and profits, time savings, and greater efficiency. As mentioned in a recent article in the *The Economist,* "mobile phones have tremendous possibilities to help boost entrepreneurship and economic activity, provide an alternative to bad roads and unreliable postal services, widen farmers' access to markets, and allow swift and secure transfers of money."[8]

Usage is probably the most critical dimension for middle-income countries as they transition into full-fledged knowledge economies. Once again, the challenge is to become and remain competitive in the global economy, using enhanced ICT as a powerful enabler. For these countries, a more developed infrastructure, deeper Internet penetration, and, perhaps most important, an increasingly demanding and sophisticated populace has led to the development of a vast array of electronic services and customized applications. Online initiatives, usually linked to e-government, typically involve delivering government services to a large percentage of the population. These help to improve government transparency and reduce the chances of corruption (particularly in the public sector), enhance the efficiency and productivity of business, and open up new opportunities such as online transactions. Brazil (box 6.8) has pioneered and mainstreamed initiatives in many of these areas.

Estonia (box 6.9), the most "wired" country of the former Soviet Union, is another example of a country that is exploiting a strong ICT infrastructure and has moved strongly in the direction of e-government.

These types of services are gradually transforming the relationship between citizens, the government, and businesses, with major economic, political, and social consequences. Middle-income countries have addressed some of the challenges of access, but they must still develop, sponsor, and support changes that will ensure that the knowledge being accessed, shared, and applied effectively delivers the tangible results needed for building a true knowledge economy.

Box 6.8 Brazil and the Internet

In 1999, the Brazilian government launched an information society program that promotes universal access, business competitiveness, and e-government. Through the universal access plan, thousands of community centers offering Internet access have been opened, many in post offices and libraries. Nearly 30 million Brazilians now use the Internet on a regular basis for many purposes, such as: voting, conducting online banking, filing taxes, and so on. The Poupatempo program consists of ICT centers that bring representatives of different government agencies together under one roof. Government workers at these centers have computers networked with each other and with their respective agencies. To receive government services, such as obtaining a birth certificate or a driver's license, a citizen need not visit different agencies; instead, he or she can go to different agents sitting in the same room. Brazil also has the distinction of being the only country with universal electronic voting during national elections. It has shared its practices in this area with Argentina, Mexico, and the Dominican Republic. In addition, business-to-business transactions using the Internet have grown significantly, municipalities are sharing knowledge through state-of-the-art videoconferencing facilities, and numerous schools have launched e-learning programs.

Source: Jeffrey Rinne, World Bank (http://unpan1.un.org/intradoc/groups/public/documents/UNPAN/UNPAN023527.pdf).

8. "Behind the Digital Divide," *The Economist,* March 10, 2005.

Box 6.9 Estonia: Applying the "e" to All Aspects of Life and Work

For the past decade, Estonia has mainstreamed ICT into all aspects of daily life. Committed political elites and the private sector stand behind the effort, and they enjoy strong public support. With universal access a national priority, the country has channeled substantial resources into expanding and improving access, particularly Internet access, and developing new services. Not surprisingly, Estonia is considered by many to be one of the most dynamic countries in Europe in terms of ICT. Its ICT infrastructure is nearly ubiquitous—personal computer penetration is at 70 percent. More than half of the population and 80 percent of businesses have direct access to the Internet (in addition to 100 percent of schools and the public sector). Eight out of ten Estonians carry a mobile phone, and even gas stations in Tallinn are equipped with WiFi connections, allowing motorists to check their e-mail after they fill up. The government and business are taking advantage of how well connected the country has become by offering a wide range of e-services. For example, most Estonians use mobile phones to pay for parking, and more than 70 percent file their taxes on the Internet. Most conduct their banking online. The government has developed more than a dozen e-government services for citizens on a sophisticated e-citizen portal, and electronic identification cards are now compulsory. Such an ICT-friendly atmosphere has spawned and attracted a large number of entrepreneurs in a society that has embraced technology as the fastest way of catching up with the West. These entrepreneurs are developing innovative and popular software such as Skype.

Source: Mark Landler, "Chilly Streets and Hot Technology in Estonia," *New York Times*, December 13, 2005.

Developing ICT Skills

Skilled human capital is a key condition for the success of ICT projects. Maintaining an ICT infrastructure and launching new ICT applications depend on an available pool of skilled workers, implying the need for technical training, general education, and the capacity to commercially exploit the knowledge that ICT makes available. A recent study of e-business has shown that the workforce skill level is one of the most significant factors affecting the export performance of Indian firms. Another study found that a one-year increase in average schooling results in a one percentage point increase in personal computer penetration.

Low-income countries must identify the most pressing ICT skills gaps and devote the resources necessary to address them. In some cases (box 6.10), they can design innovative programs that integrate a capacity-building component. In others, partnerships can be forged with donors or the private sector. In July 2006, Microsoft pledged to provide computer training to more than 45 million Africans by 2010. The goal of the program, which is similar to projects that the U.S.-based software company has funded in Mexico and Chile, is to bring Internet technology to more than 600,000 schools across Africa.

In some low-income countries—notably China, India, and other countries in East Asia—the ICT sector has proven to be a significant generator of employment and economic growth. Through the application of modern technologies, coupled with increases in productivity, Vietnam's ICT sector has quickly gained in competitiveness. Hardware grew by 27 percent between 2002 and 2003, and software by 40 percent. By the end of 2003, there were about 2,500 registered ICT businesses—

Box 6.10 *Developing the IT Skills of Poor Women in Kerala*

Traditionally, governments have developed a large in-house IT function or have outsourced their purchasing of IT goods and services to the private sector—often to multinational subsidiaries. Now, there is a third way: outsourcing data entry, digitization, hardware/software purchase, IT training, computer servicing and maintenance, and so on to social enterprises such as cooperative IT enterprises created by poor women. An innovative program in India is producing a very positive outcome that goes beyond its initial ICT focus.

The program contracts e-government work to social enterprises in the state of Kerala. Within its poverty reduction strategy, Kerala has fully integrated its e-government program, introducing a large number of computers and software applications to government employees for the first time. The plan has involved outsourcing the work of installing computers, data entry, and IT training to women trained in advance through cooperatives. The results have been threefold: a cost savings to the state, a reduction of poverty, and the image enhancement of the women involved. Kerala's initiative with women's IT enterprises shows what can be achieved: from a standing start five years ago, it now has more than 200 such enterprises employing 4,000 previously unemployed women from impoverished families.

Source: Heeks and Arun 2006.

including 570 software companies—and 12,000 ICT workers. But challenges remain. Vietnam's IT workforce is small, inexperienced, and inadequate. Of 40 million workers in Vietnam, only 20,000 are in IT, and only 3,500–4,000 students graduate with IT qualifications every year. To be globally competitive, Vietnam needs more experienced IT professionals (World Bank 2005f; Chidamber 2003). A shortage of IT graduates may be even more of a problem for India, where a shortfall of 500,000 trained IT professionals has been forecasted.[9] Still, most low-income countries lack the resources and capacity to develop their own IT industry. Instead, they should concentrate on developing basic skills and literacy in ICT.

Owing to their generally stronger education systems, middle-income countries usually have a more literate population as well as a significant number of well-trained professionals in both the public and private sectors. However, these countries cannot become complacent. As public access becomes widespread and even universal, and as important services become available through the Internet and mobile phones, the population as a whole requires a certain level of ICT literacy. Enhanced skills will need to be more widely distributed. This has implications for education systems and specialized training programs. Middle-income countries must continue to promote and support policies that ensure sufficient human capacity in ICT.

Jordan has made skills development a priority as part of its goal to become a regional center of ICT and, more importantly, a knowledge economy. Its information and communications technology initiative supports a national vision for realizing the potential of ICT through a set of actions that includes upgrading the skills of people across the country so that they can better use and develop new technologies.

9. For more details, see "If in Doubt, Farm It Out," *The Economist,* June 1, 2006.

The initiative has stimulated a number of public–private partnerships between government ministries and companies such as Cisco and Microsoft, which are improving ICT skills development at all levels of education and enhancing the country's capacity in networking, technology management, and IT services—high-end skills that can be leveraged regionally.

Evolving Agendas

The growth of ICT in many countries can be divided into three stages that reflect a gradual effort to make ICT into a genuine instrument of development and economic growth. The "i-agenda" of the 1980s and early 1990s centered on governments' efforts to achieve greater connectivity and fulfill the requisite conditions for ICT-based growth. Starting in the mid-1990s, governments focused on the "e-agenda," with efforts in specific sectors such as health and education. Over the past few years, a vision combining the two earlier agendas has emerged in a number of countries, a vision that highlights the ability to mobilize and use knowledge to determine a nation's role in the global economy: the "k-agenda." As stated by Lanvin and Neto (2005): "the e-agenda has become a core set of objectives and priorities within a larger knowledge agenda and ceases to be an end in itself."

It would be easy to create a dichotomy and suggest that low- and middle-income countries are at different stages in the building of a strong ICT infrastructure and therefore face different challenges and opportunities. But that distinction does not always hold and it may obscure what is really taking place. For example, some low-income countries may be in a better position to adopt newer technologies than middle-income ones, where entrenched interests and less flexible governments often represent a formidable obstacle. Geographic considerations such as country size also matter, as witnessed by the relative success of island countries in building a strong ICT infrastructure. Further, all countries, regardless of income level, are in a position to learn from and incorporate the experiences of more developed countries, which have mainstreamed ICT into all aspects of their economies. Not surprisingly, the business model for developing ICT will—and should—differ from country to country. It should be based on each country's economic, social, and political conditions, as well as its technical infrastructure and know-how. Finally, it should be based on tackling the fundamental question of how to use ICT to foster productivity and innovation, stimulate growth and prosperity, increase competitiveness in global markets, and help countries become genuine knowledge economies.

7

Upgrading Education

Education creates choices and opportunities, reduces poverty, and gives people a stronger voice in society. It is the fundamental enabler of the knowledge economy. Well-educated and skilled people are essential for creating, sharing, disseminating, and using knowledge effectively in a global environment that is radically changing the types of skills needed for economic success.

Basic education provides the foundation for lifelong learning and increases people's capacity to assimilate and use information. Secondary and tertiary education should develop core skills (including technical skills) that encourage the creative and critical thinking inherent in problem solving and innovation. Higher education in engineering and the sciences is needed to monitor technological trends and to use new technologies while assessing which are relevant for a particular firm or the economy in general. The production of new knowledge and its adaptation to a particular economic setting is generally associated with higher-level education and research.

Indeed, a better and more broadly educated population tends to be more technologically sophisticated, thus generating quality-sensitive demand for advanced goods. This, in turn, tends to stimulate local firms to innovate—to design technologically sophisticated goods and adopt advanced production techniques.

A culture of continuous learning and openness to new ideas is critical for a knowledge-based economy. From early childhood through retirement, a lifelong learning system encompasses learning from all types of sources: formal (schools, training institutions, and universities), nonformal (on the job and in the household), and informal (family or community members). The basic requirements of a lifelong learning system are comprehensiveness, multiple pathways to learning and problem solving, and multiple education providers. To benefit from such a system, recipients must have the ability to acquire new skills, act autonomously, use tools interactively, and function in socially heterogeneous groups. Basic competencies required by employers in today's labor markets (table 7.1) give an indication of the challenges faced by education systems.

Like policies related to information and communication technology (ICT), efforts to build a knowledge economy through education need to be adapted to each country's economic and social needs as well as to its institutional capabilities. These will vary depending not only on the level of development but also on political, cultural, and other country-specific factors. Table 7.2 summarizes priorities for both low- and middle-income countries, and relevant policies at different levels of education.

The following sections (a) emphasize key points to keep in mind when adjusting and reforming education systems to meet the requirements of the knowledge

Table 7.1 *New Competencies for the Knowledge Economy*

Competency area	*Content*
Cognitive	Language, communication, logistical and mathematical thought
Problem solving	Observing, analyzing, identifying the parts of a problem, suggesting creative solutions, critical thinking, planning and project management skills
	Adapting knowledge to new contexts
Self-learning and self-knowledge	Being informed and motivated to learn, concern with one's own development, knowledge of one's capacities, ability to transfer knowledge from one context to another
Social	Working in a team, negotiating and creating constructive arguments, interacting, getting others to understand one's point of view, self-confidence, seeking and maintaining networks of social contacts
Motivation for work	Initiative, responsibility, commitment, and interest

Source: Adapted from Vargas Zuñiga 2005.

Table 7.2 *Priorities for Educational Policies: An Overview*

Low-income countries

Primary education	*Secondary education*	*Tertiary education*	*Lifelong learning*
Focus on Education for All (EFA) targets. Expand enrollment. Increase the education and employability of women. Focus on quality.	Increase lower-secondary enrollment. Develop upper-secondary programs relevant for working life as well as for tertiary education.	Improve quality of tertiary education. Align with skill needs of economy. Get employers to identify program and skill needs. Develop shorter-duration technical education programs.	Adult education: Develop campaigns and provisions to increase literacy. Provide training in skills required by economy. Provide skill upgrading.

Middle-income countries

Primary education	*Secondary education*	*Tertiary education*	*Lifelong learning*
Improve quality, focus on completion. Target low enrollment rates in poorer areas. Reduce number of repeat students. Improve quality and relevance of teaching.	Aim for universal coverage in lower-secondary programs. Increase enrollment rates in upper-secondary programs. Improve quality and relevance for employment and life skills. Develop permeability from vocational to general track and to tertiary education.	Aim to increase tertiary enrollment. Ensure that disciplines and curricula are what the economy requires. Strengthen specialized short-term degree programs. Promote different types of providers. Aim for international qualifications and transferability.	Develop credit-transfer systems. Strengthen qualifications for employment. Improve quality assurance and assessment mechanisms across entire educational and training system. Strengthen system governance through multiple pathways and providers.

Source: Authors.

economy, (b) illustrate policy principles with pertinent examples, and (c) show how various ICTs can help to improve and scale-up conditions for delivering education and training.

Basic Education

Primary Education

The focus of low-income countries should be on achieving basic education for all, as articulated in the international Education for All (EFA) agenda (box 7.1). Although significant progress has been made in increasing enrollment, policies should now focus on efficiency, quality, and equity issues such as narrowing the gender gap in economic opportunity and the rural-urban divide.

At the level of basic education, this implies a move from quantitative enrollment to enhanced quality and skills. Box 7.2 illustrates the example of Ghana, which has both increased enrollment and improved learning outcomes.

Most low-income countries currently suffer from various problems directly relevant to their education systems, among which are a low percentage of sufficiently educated people (40 percent illiteracy rate on average), inequitable and inefficient education policies, lack of curricula and teaching methods relevant to the country's economic context, and the general weakness of science and technology training.

Other obstacles to overcome are the scientific and digital divide, poverty and scarcity of resources, and the gap between the language used for teaching and that spoken every day. This last point is essential, as it affects the durability and depth of knowledge acquired by children in school. Basic knowledge and competence are best acquired by learning in the local language in the first years, and gradually learning a second language. After some years, the foreign language may become the first. This approach helps root the learning process in a solid language base before integrating youth into the broader world (box 7.3).

Box 7.1 The Education for All–Fast Track Initiative

World leaders have recognized that it is impossible to achieve one of the Millennium Development Goals—free, universal primary education for all by 2015—without significant further effort. That recognition motivated the creation of the Education for All–Fast Track Initiative (FTI) in 2002.

Based on mutual accountability, the FTI and aims to provide incentives and resources to empower low-income nations to develop and implement sound education plans. More than 30 bilateral, regional, and international agencies and development banks are providing greater, and better coordinated, financial and technical support. As of November 2006, 28 developing countries had joined the FTI by agreeing to put primary education at the forefront of their domestic efforts and to develop sound national education plans. Through the initiative, all involved partner countries and agencies coordinate at the national and international levels to ensure greater donor cooperation, knowledge sharing, and resource mobilization.

The UNESCO EFA Global Monitoring Report 2007 (2006) estimates that reaching the Universal Primary Completion goal in low-income countries will require an increase in aid for basic education to $9 billion per year (from current levels of $2.7 billion per year). Potentially 32 FTI countries, including India, could join the EFA initiative by the end of 2008.

Source: EFA-FTI Secretariat.

Box 7.2 Higher Enrollment and Better Learning Outcomes in Ghana

Today, nearly 100 million of the world's children do not go to school, and, of that number, more than 70 percent live in Sub-Saharan Africa and southern Asia. More than 15 percent of children in developing countries do not complete a course of primary education; in Sub-Saharan Africa, the figure exceeds 40 percent. Over 50 percent of African school graduates cannot read. Remedying this situation is an immense challenge, but Ghana has already demonstrated that it is possible to increase enrollment while improving learning outcomes.

In 2000, more than 90 percent of Ghanaians aged 15 and above had attended school, in contrast to 75 percent just 20 years earlier. In 1987, nearly two-thirds of primary school graduates were unable to read. Today, fewer than one in 20 is illiterate. The quantity and quality of schooling in Ghana has improved as a result of the authorities' commitment to a program of educational reform that has refocused resources away from elitist secondary and tertiary education toward primary education. World Bank assistance was important in supporting the policy reform process and in financing school-level quality improvements such as classroom construction and rehabilitation and the widespread availability of teaching materials. A report by the World Bank's Independent Evaluation Group demonstrated that Bank-financed textbook provision accounted for around one-quarter of the observed improvement in test scores. Nevertheless, the report signaled important gaps: "Even in countries where learning outcomes have improved, absolute levels of student achievement are still low . . . In Ghana only 10 percent of children reached the country's mastery levels in math and 5 percent in English. So, even though there has been progress, other school-level determinants of achievement such as the professional support of teachers and improved teaching methods still need increased attention in Ghana."

Source: Adapted from World Bank 2006c.

Box 7.3 Multilingual Education in Africa

The UNESCO Institute for Education (UIE) and the Association for the Development of Education in Africa (ADEA) have carried out a study of multilingual education that aims to help policy makers and educators make informed decisions about language use in education. Such decisions determine not only the rate of social return on investments in education but also that of the development of society at large. In 2005, a review of experience and educational programs using mother-tongue and bilingual education was undertaken in 25 Sub-Saharan African countries. The researchers analyzed practices and issues in six thematic areas: policies and management, teaching and learning practices in both formal and informal education, language education models, costs, and publishing. The study clarified the difference between learning a language and learning through a language. Because implementing mother-tongue and bilingual education is often perceived as too costly, the study analyzed the real costs. The findings clearly show that the most beneficial and effective education system in multilingual African nations is "additive multilingual education" that leads to a high proficiency in both African and international languages.

Source: Adapted from ADEA 2005.

Secondary Education

As the EFA initiative progresses, an increase in the demand for secondary schooling is very likely. Indeed, a surge has already taken place in a large number of low- and middle-income countries (figure 7.1). It will require policy makers to expand access to disadvantaged groups; improve labor market relevance; and address teacher quality, governance, and financing issues in secondary education. In Sub-Saharan and North Africa, the Middle East, Latin America, and Asia, secondary education reform is becoming an integral part of EFA efforts (World Bank 2005e).

Investment in secondary education yields both individual and social returns and provides many of the skills and knowledge needed for countries' economic growth, including professional training such as that required by technicians, scientists, and entrepreneurs. It also helps socialize young people, including at-risk youth. Following the rapid progress made in primary-level enrollment under the EFA initiative, demand for secondary education has been rising rapidly in low-income countries, especially at the lower-secondary level. In many nations, however, much of the secondary curriculum remains ill-adapted to the current international and national context and is in dire need of reform.

Central to the question of reform is the growing relevance of secondary education to the needs and means of poor countries—including the type of education and pedagogy proposed, the professional development of teachers, and the ICTs used in delivery.

Figure 7.1 Trends in Gross Secondary Enrollment from 1995 to Most Recent Year

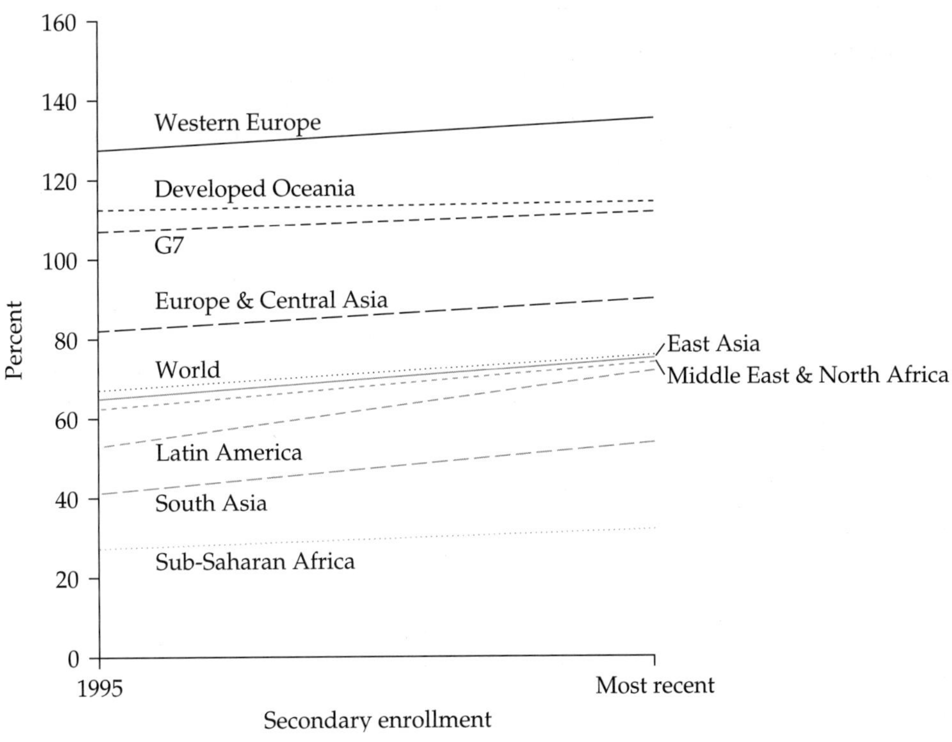

Source: Authors' calculations from KAM version 2006.
Note: The "most recent year" is generally 2004–05.

The costs of secondary education are high. In view of the severe financial constraints faced by low-income countries, access will not be expanded without major changes in the way secondary education is delivered. These include modifications in structure and organization, improved teacher management, and alternative modes of delivery and financing.

Two examples of middle-income countries that have implemented such changes are given below. The first, Colombia (box 7.4), reached the rural poor through distance learning. The second, Jordan (box 7.5), teaches youth with innovative methods delivered through public-private partnerships among Jordanian education authorities, private hardware and software companies, and international donors.

Vocational Training

Vocational training is of crucial importance to building the basic skills needed in low- and middle-income countries. The competence level of a country's skilled workers and technicians is centrally important to the flexibility and productivity of its labor force.

Evaluations performed by the World Bank (De Ferranti and others 2003) criticize the limits of vocational training policies and programs to improve the competitiveness and growth rate of industries in the developing world. Serious adjustments are needed. Fortunately, significant improvements can be made through true decentralization of program management (table 7.3). The movement to provide individual training institutions and managers with increased autonomy and accountability

Box 7.4 *Colombia's Telesecundaria: Reaching the Rural Poor*

Telesecundaria in Colombia (originally developed and implemented in Mexico) expands rural access to quality secondary education. It is an educative, pedagogical, and didactic model for addressing the needs of rural young people between the ages of 12 and 17 who have finished two or three years of elementary education but have been unable to continue to secondary education. Telesecundaria aims to be a comprehensive instructional model that enables schools to deliver a complete lower-secondary curriculum at costs comparable to those provided in more populated, urban areas.

The program, based on remote broadcasting, provides teachers and students in rural areas with a complete support package that includes a combination of written and audiovisual materials (regularly evaluated and updated) and specific training modules for teachers. Video recordings (including more than 130 newly developed videos on Colombian history, geography, ethics, and human values) and satellite broadcasts have been successful in answering students' educational needs. The pedagogical strategy of Telesecundaria is active and cooperative learning through teamwork, promoted through classroom research and community development.

Telesecundaria has achieved wide recognition in its few years of existence, and 70 percent of rural Colombian communities now consider it very important. The main benefit is the opportunity it gives young people who have finished elementary education to continue their education. A corollary benefit is community integration: students participate in community activities, creating cohesion for other projects that benefit the entire community.

Source: World Bank 2000d.

Box 7.5 The Jordan Education Initiative: New Methods for Improved Teaching

The Jordan Education Initiative (JEI) grew out of an extraordinary challenge, posed at the 2003 World Economic Forum, to accelerate social and economic development in Jordan through the broad application of e-learning hardware, curricula, and training. The ultimate goal of the acceleration was to produce sustainable economic growth and support the development of a knowledge economy. JEI has four main objectives: to improve the delivery of education in Jordan through public-private partnerships; to foster innovation among teachers and students through the effective use of ICT; to build the capacity of the local information technology industry; and to create a model of reform that can be used by other countries. Program implementation began with the design, development, and deployment of a mathematics curriculum for grades K–12 to be delivered in more than 100 schools. The multimedia content is Web-based and interactive.

A McKinsey study shows that, by 2005, JEI had more than 30 active partners from public sector organizations such as USAID and the British Council, and private sector companies including Cisco, Dell, and Microsoft. One hundred discovery schools are now fully networked, with access to computer labs and online curricula. A full math e-curriculum (grades 1–12) has been developed; Arabic online, English as a foreign language, and ICT are being tested at varying stages in the discovery schools; and online civics and science courses are being developed.

Although it is too early to judge the long-term impact, there are early indications of positive outcomes. Informal assessments by external educational experts and members of the Jordanian ministry of education show that results are promising. The initial stage of the JEI e-curricula and technology has benefited approximately 2,300 teachers and 50,000 students in the 100 discovery schools. More remains to be done, however, to train teachers and principals and to deploy technology and e-curricula more broadly. Over the long term, the intention is to roll out the JEI learning model to all public schools in Jordan.

A second intended outcome of the JEI is to develop the Jordanian ICT industry by encouraging partnerships with global firms. Five local companies have benefited from close working relationships with global partners, and McKinsey estimates that approximately $3.7 million has been transferred to local companies as a direct result of JEI programs.

A third outcome is already in evidence: the Jordanian government has received inquiries from more than 10 countries interested in adapting elements of the JEI model for education reform, and the World Economic Forum is exploring efforts to export the model to the Palestinian territories, the Arab Republic of Egypt, and India.

Source: McKinsey and Company 2005.

for results has improved the relevance and quality of skill development. Centralized public training systems have become more responsive where individual training institutes have been given the freedom to set fees, adapt training to local needs, hire appropriate staff, and choose methods of instruction. The shift from financing inputs for training to financing performance and outcomes has helped change incentives for improvements.

As illustrated in table 7.3, government ministries responsible for vocational education should decentralize many functions to training institutions and concentrate on (a) developing policies, setting standards, and improving public information about the vocational education and training system; (b) financing training

Table 7.3 Achieving a KE Advantage in Vocational Education through Decentralization

Example of centralized vocational education model		
Ministry	*Training institution*	
• strategy • standards • generic relation with private sector • staffing • curricula • budget allocation • investments • student intake	• administrative duties • delivery of training courses	
Example of decentralized vocational education model		
Ministry	*Training institution*	*Enterprises*
• strategy • standards • generic relation with private sector • budget allocation • investments in priority areas • evaluation	• administrative duties, staffing • delivery of training courses • student intake • curriculum development • partnership with private sector • monitoring	• participation in course delivery • participation in curriculum development • cofinancing • testing

Source: Authors.

to meet equity objectives and to fill strategic skills gaps; and (c) providing skills training in priority areas where nongovernment providers are reluctant to invest (while exercising caution to avoid crowding out such providers). Furthermore, nongovernment providers should, if possible, participate in course delivery and curriculum development as well as function as cosponsors to ensure the relevance of training activities. Programs may have to be redesigned to address the needs of local communities in poor areas that lack the basic skills for adopting and repairing technologies (see box 5.7 on Rwanda).

Box 7.6 illustrates the special issues that are important for building skills for the informal economy. Though training through traditional apprenticeships is self-financed, self-regulated, and cost effective, it perpetuates traditional technologies and lacks standards and quality assurance. The informal sector is where most of the nonfarming poor work, and where investments in skills development—along with complementary inputs such as access to secure workplaces, credit, and technology—can play an important role in poverty reduction, particularly for women and other vulnerable groups.

Tertiary Education

Improving Quantity and Quality through an Enabling Framework and Appropriate Incentives

Knowledge accumulation and application have become major factors in economic development and are increasingly at the core of national competitive advantage in

Box 7.6 Vocational Education and Training in the Informal Sector

Demand for training among small enterprises in the informal sector is likely to be low in the absence of clear demonstration of the benefits of skills development. Informal sector associations can be helpful in raising the awareness of skill shortages among members, including literacy. Successful examples of programs that combine learning for livelihoods with literacy training are available.

Training for the informal sector is necessarily different from that for the formal sector in its preference for merging technical and business management skills and delivering the resulting courses on a flexible schedule. Poor people have shown a willingness to pay for good training, but it must have immediate application, since the poor can hardly afford long periods of training before seeing a payoff. Full cost recovery is rarely attempted or achieved by such programs, although subsidies can be justified on grounds of social equity and efficiency. Focusing on the evaluation of competencies achieved through training is important to quality assurance.

Source: Adapted from Adams and Johanson 2004.

the global economy.[1] The role of tertiary education in the construction of knowledge economies is crucial. Many developing countries, however, continue to wrestle with difficulties stemming from unresolved problems such as expanding education coverage in a sustainable way, inequalities of access and outcomes, educational quality and relevance, and inflexible governance structures and management practices.

There is no blueprint for the way all countries should reform tertiary-education systems, but a common prerequisite seems to be a clear plan for the long-term development of a comprehensive, diversified, and well-articulated system. Student mobility can be encouraged by developing open systems that (a) recognize relevant prior experience, degree equivalency, credit transfer, and tuition exchange schemes; (b) provide access to scholarships and student loans; and (c) establish comprehensive qualifications and a lifelong learning framework. The regulatory environment should be one that encourages rather than stifles the innovations of public institutions or private sector initiatives to expand access to good-quality tertiary education. New funding for tertiary education should be channeled to diverse institutions, including private providers (Salmi 2004).

There is a worldwide trend toward organizing the higher education system along the following pattern: undergraduate (three to four years of study following the completion of secondary education), master's (five years of postsecondary education), and doctorate (three years of study after the master's). This three-tier system has proven its effectiveness for building needed competencies ranging from advanced education to research. It is being generalized in Europe and Asia and adapted to low-income countries (such as those in francophone Africa).

The first and primary goal for low-income countries is to expand the number and quality of graduates, thus creating a base on which to build further efforts. Experience shows that this goal is possible to achieve if appropriately planned and supported by strong vision and leadership, as in Mozambique (box 7.7).

---

1. For an extensive discussion of the role of higher education in the construction of knowledge societies, see Salmi (2004).

Box 7.7 Mozambique's Tertiary Education Reforms

Mozambique is an example of a country that has taken steps to increase the quantity and quality of its university graduates. In 1990, there were fewer than 3,000 university-trained people in the nation. Fewer than 15 percent of civil service personnel had a university degree, and more than a third of this group had less than seven years of primary education. Highly trained and professionally experienced Mozambicans tended to go into the private sector—with adverse effects on the planning and management of both public and private sector activities. However, as a result of a World Bank–assisted human resources development project implemented between 1993 and 2001, in which ministry structures were the stakeholders, the number of students graduating annually from Eduardo Mondlane University increased nearly five times—from 148 in 1994 to 695 in 2000. In addition, the percentage of graduates who completed all degree requirements within the officially prescribed time increased from 5 percent in 1997 to 15 percent in 2000, and the overall average examination pass rate rose from 39 percent in 1994 to 53 percent in 2000. Critical capacity-building needs were addressed through a strategy based on links with selected foreign universities. The proportion of the university's Mozambican teaching staff with graduate degrees increased from 17 percent in 1994 to 46 percent (exceeding the target of 32 percent) in 2000.

Source: World Bank 2003f.

Low- and middle-income countries necessarily encounter problems of distribution and quality. Mexico's Monterrey Institute of Technology provides an exceptional example of a way to tackle such issues (box 7.8).

The University's "Third Mission"

Tertiary-level and research institutions in low-income countries need to focus on creating a pool of experts capable of adopting science and technology and adapting it to the local context. In particular, this means changing the current paradigm to include—in addition to teaching and research—a third mission: service to the community and close cooperation with the public and private sectors to contribute to innovation and development. (An example of how this was done by an African institute studying malaria is described in box 5.6.) It is imperative that universities work toward fulfilling their mission of community service—a mission that has been promoted in industrialized countries over the past two to three decades.

Another necessary change in the current academic paradigm, also illustrated by the example mentioned above, is the need to tap into global knowledge by creating regional networks and communities of practice, poles of excellence, and both South-South and North-South partnerships between institutions. Finally, given resource constraints, tertiary-education and research institutions need to undertake managerial and financing reform to reinforce their autonomy and competitiveness.

In sum, at the tertiary level, policy needs to be pragmatic and flexible to help build a qualified labor force that is able to adapt to changing demand in both local and global markets. This implies a shift from public financing to public-private partnerships and the implementation of (a) curricula that include the knowledge and skills required for the new economic context, (b) new mixed-mode teaching to complement traditional face-to-face teaching, (c) and more open policies toward society and the private sector, as opposed to traditional "ivory tower" practices (table 7.4).

Box 7.8 Millennium University of the Monterrey Institute of Technology in Mexico

The Monterrey Institute of Technology in Mexico is a premier private education organization comprising a network of 33 campuses throughout the country. It is a franchise system of local campuses, each of which is financed and governed by local private sector leaders. Its virtual university is a leader in distance learning, championing a continuing education agenda throughout the Spanish-speaking world and making inroads into other giant markets such as China.

To reach students with few financial resources, the institute launched a spin-off, TecMilenio (Millennium University), designed to combine the high-quality education associated with the institute brand with dramatically lower costs. Four years after its founding, TecMilenio has 33 campuses in Mexico and an enrollment of more than 17,000 students—with per student costs approximately three times lower than in the parent organization. Factors that have allowed it to dramatically reduce costs without compromising quality include:

- A curriculum designed and often delivered by the management of private sector firms. TecMilenio shares offices with some of these firms so that students and teachers often work, learn, and teach in the same location.
- The utilization of distance education, which gives students access to the best professors and courses.
- Pedagogy based on problem solving, conceptual tests, and standardized and centralized testing.
- Teacher remuneration dependent on the test results of students. A small management structure draws on carefully selected professors from institute staff, and translates industry needs into pragmatic curricula, curbing the tendency of professors to draw on the same teaching materials they have been using for decades.
- Facilities that are used for teaching only; there are no cultural activities or sports facilities, which allows for lower tuition costs. By the year 2010, TecMilenio plans to reach an enrollment of 100,000 students.

Source: Authors.

Table 7.4 Higher Education Policy Requirements for the Knowledge Economy

Traditional policy orientations	*Knowledge economy requirements*
• Public financing	• Public-private partnership
• Traditionally separate disciplines	• New knowledge adapted to social and economic requirements
• Face-to-face teaching	• Mixed modes with distance and e-learning components
• Inward orientation focused on academic communities	• Opening to local communities and the business sector
• National standards and cooperation with national institutions	• Inclusion in global networks of higher education institutions, with access to related performance evaluations

Source: Authors.

Lifelong Learning

In the transition to a knowledge economy, an efficient, nationwide, lifelong learning system is an important building block for all countries at all levels of development. The challenge is to establish systems based on learning needs rather than on age, and to replace information-based, rote learning with educational practices that develop a learner's ability to learn, create, adapt, and apply knowledge.

General Principles

In developing countries, five main policy actions are suggested for pursuing a lifelong learning strategy (OECD 2004b):

- Improving access, quality, and equity throughout the formal education system
- Ensuring foundation skills for all
- Recognizing all forms of learning, not just formal courses of study
- Mobilizing resources and rethinking resource allocation across all sectors, settings, and over the life cycle
- Ensuring collaboration among a wide range of partners.

In addition, building lifelong systems requires the systematic codification of knowledge (into categories such as general, technical, and professional). This implies segmenting knowledge into modules, or credit hours. For technical and vocational knowledge, codification should be carried out on the basis of competence, qualification, and vocational skills (and the connections among them) so as to promote cultural assets that are socially and professionally recognized and can be used by the individual in accordance with his or her experience. The necessary codification requires serious reflection about the nature of knowledge itself in order to define professional profiles and related skills, modularize general education (to favor the earning of qualifications at different stages), develop instruments at different levels for the validation or accreditation of technical and vocational skills, and reexamine technical and vocational paths and possibilities for general education.

While lifelong learning encourages multiple providers and pathways, it also requires better coordination among government and private sector stakeholders; better assessment, accreditation, certification, and vocational qualification systems; and better information services (Dahlman, Zeng, and Wang 2007). The World Bank has produced a report with recommendations on how China can build a comprehensive lifelong learning system (see box 8.1).

Adult Education

Improving adult education, as well as upgrading the skills of workers, is an important issue for low-income countries. Indeed, increasing the adult literacy rate, as well as providing the necessary training in skills required by the economy, is essential. Again the use of ITCs can be very useful in this respect.[2]

Adapting the labor force to changing market demands also requires appropriate policies. Policy instruments to support training include cofinancing arrangements, tax incentives, and subsidies. It is important that low- and middle-income countries have stable training programs that respond to business demand—not least in their growing export industries—coupled with evaluations to favor those schemes with a proven track record of high rates of return. Television-based programs can be very helpful, as has been found in Brazil, India, and Mexico (box 7.9).

2. As illustrated by an initiative in India, the Computer-Based Functional Literacy (CBFL) program (see www.tataliteracy.com).

Box 7.9 *Use of Educational Television in India, Brazil, and Mexico*

Two educational television programs in Latin America and one in Asia—Mexico's Telesecundaria, Brazil's Telecurso, and the National Open School of India—have succeeded in providing secondary education to students who would not otherwise have received it. The success of television-based education programs in these three countries, despite the high initial fixed-cost investments, is attributed to their large populations of potential secondary-school students, which permit economies of scale.

Mexico's Telesecundaria was developed in 1968 without external financing. Its main objective was to solve the problem of access in rural areas. (The urban delivery version became financially unviable because of low demand.) Telesecundaria targeted students in the 200,000 rural communities with populations of less than 2,500. In 1998, 15 percent of Mexico's lower-secondary students were educated through the program.

Brazil's Telecurso was developed in the late 1970s by the Roberto Marinho Foundation and supported by the country's largest commercial network, Rede Globo. Its purpose was to provide young working adults with the opportunity to acquire primary or secondary equivalency certificates. To respond to new labor market demands, a new program, Telecurso 2000, was developed in the early 1990s by the Roberto Marinho Foundation, the Globo media company, and private sector investors. Telecurso 2000 is a condensed version of a basic secondary education curriculum, with an optional curriculum on basic mechanical skills. The guiding principles of its design are job-oriented education, development of basic skills, citizenship education, and contextualization.

The National Open School of India was established in 1989 as an autonomous institution under the Ministry of Human Resources Development to support India's National Policy on Education. The school caters to the educational needs of out-of-school children and of socially and economically disadvantaged students. Although it started with academic courses at the secondary level (including the upper-secondary level), it currently offers courses in vocational and other life-skills areas. It has also extended its range from elementary to preuniversity programs. More than 400,000 children from physically, socially, economically, and geographically disadvantaged groups have enrolled in the school.

Source: World Bank 2005e.

Governance of Education Systems

Given the complex set of challenges facing the education sector, policy makers and educators need to put the individual learner and his or her demands at the center of education policies for the knowledge economy.

Many governments, and not only those of developing countries, are struggling to change the governance of their education systems. Education policies tend to be strongly sectoral, where different education sectors (such as primary, general and vocational, secondary and tertiary) compete for public resources. This often makes it difficult to move public resources from one education sector to another and to have sectors collaborate together. To get away from traditional supply-side offerings requires a strong and sustainable political commitment to lifelong learning and a systemic approach that views the demand for and supply of learning opportunities as part of a connected system that covers the whole life cycle, including all forms of formal and informal learning. It is also becoming increasingly important to coordinate education policies with labor market, research, and innovation policies. Aspects of the needed changes in education governance are summarized in table 7.5.

Table 7.5 The Traditional versus the New Role of Education Governance in the Knowledge Economy

Policy issue	*Traditional role*	*New role in the knowledge economy*
• Integration/coordination at national level	• Adopts compartmentalized, sectoral approach	• Coordinates multisectoral approach
• Coordination across governance levels	• One-way control and regulation	• Two-way mutual support and partnerships
• Government as an enabler	• Controls and regulates	• Creates choices, provides information and incentives, facilitates cooperation and provision
• Link between education and the labor market/society	• Supply is institution driven	• Demand is learner driven
• Qualification-assurance system	• National standards linked with curriculum and student assessment	• Diverse system of recognition and quality control
• Administration and management	• Provides rules and regulation	• Creates incentives, facilitates diverse providers

Source: World Bank 2002a.

Autonomy and Accountability

Governance is an issue for education and learning systems in countries at all income levels. Schools and teachers cannot be made accountable for results unless they have sufficient autonomy, resources, and opportunities to build capacity. Conversely, schools cannot be given autonomy unless they have clear objectives and financing, and their progress is regularly assessed. These issues have been illustrated above with regard to vocational training.

Studies conducted by the OECD Program for International Student Assessment (PISA) have shown that 50 percent or more of the variance in performance across schools is due to differences in students' socioeconomic status—a factor that generally is not under schools' control (though, in poorer countries, schools have a larger effect). It is important to identify and agree on outcome and progress indicators based on monitoring and evaluation systems. Educational strategies and policies are all too often supply driven and not sufficiently results based. Accountability and incentive measures in education systems need to be transparent and foster progress. This means that educational outcomes should be analyzed and their implications agreed upon by policy makers, school authorities, teachers, and parents.

Box 7.10 presents the results from the PISA assessment of student learning, which shows the need for decentralizing authority to the school level while increasing the accountability of schools and teachers. Under the right circumstances, overall student performance will increase as a result.

The Third International Mathematics and Science Study (TIMSS) has created a database of student performance indicators from more than 266,000 students in the schooling systems of 29 countries. It reaches similar conclusions: schools should be allowed to decide on operational tasks autonomously, must be held accountable, and should provide teacher incentives based on improved performance.

> ***Box 7.10** School Management and Student Performance: Results from PISA*
>
> Some 315,000 students in 43 countries participated in a 2000 survey by the OECD Program for International Student Assessment (PISA) that assessed the knowledge, skills, and other characteristics of 15-year-old students. The results give some indication of the features of school systems, as well as teacher and classroom practices that affect student outcomes. Three particular observations emerged:
>
> (a) Successful educational systems are increasing schools' autonomy, aiming to raise performance levels by devolving responsibility for issues such as the choice of courses to the local level.
>
> (b) Successful school systems monitor student and system performance. Higher-performing countries have put increased emphasis on the monitoring of their schooling systems.
>
> (c) High teacher morale and expectations, as well as the use of student assessment and disciplinary measures in the classroom, are strongly related to student performance.
>
> *Source:* OECD 2004c.

Education and Equity

A major issue for developing countries is the need to maintain equity across school systems and to ensure a balanced delivery of education despite socioeconomic disparities. The experience of developed countries shows that this is possible. The 2000 PISA survey showed that high performance standards need not be achieved at the price of increased disparity.

Figure 7.2 shows countries' average performance in literacy levels (vertical axis) and social inequalities in the distribution of learning opportunities, as measured by the strength of the relationship between social background and students' knowledge and skills (horizontal axis). In particular, from the examples of Canada, Finland, Japan, the Republic of Korea, and Sweden it appears that a disadvantaged socioeconomic background does not automatically correlate with poor performance in school.

International Mobility of Human Resources

The movement of highly skilled workers and students across borders is a real and complex phenomenon that poses challenges for developing countries—with no simple solution.

Student mobility represents by far the most important source of international trade in higher education. In 2003, the United States received the greatest number of foreign students (in absolute terms) with 28 percent, followed by the United Kingdom (12 percent), Germany (11 percent), France (10 percent), and Australia (9 percent). China, the Republic of Korea, and India top the list of countries sending students outside national borders for the purpose of higher education. Asia alone accounts for 47 percent of all international tertiary-level students in the OECD area, while Africa represents just 12 percent. Receiving countries benefit from a larger pool of talent, greater scientific and technological skills, and enhanced entrepreneurial capacities.

But the key issue is whether developing countries in fact benefit from sending their best students and most highly skilled workers abroad. If these students later return to their home countries, their training can be of significant value. All too

Figure 7.2 Some Countries Achieve Both High Performance Standards and Equitable Distribution of Learning Outcomes

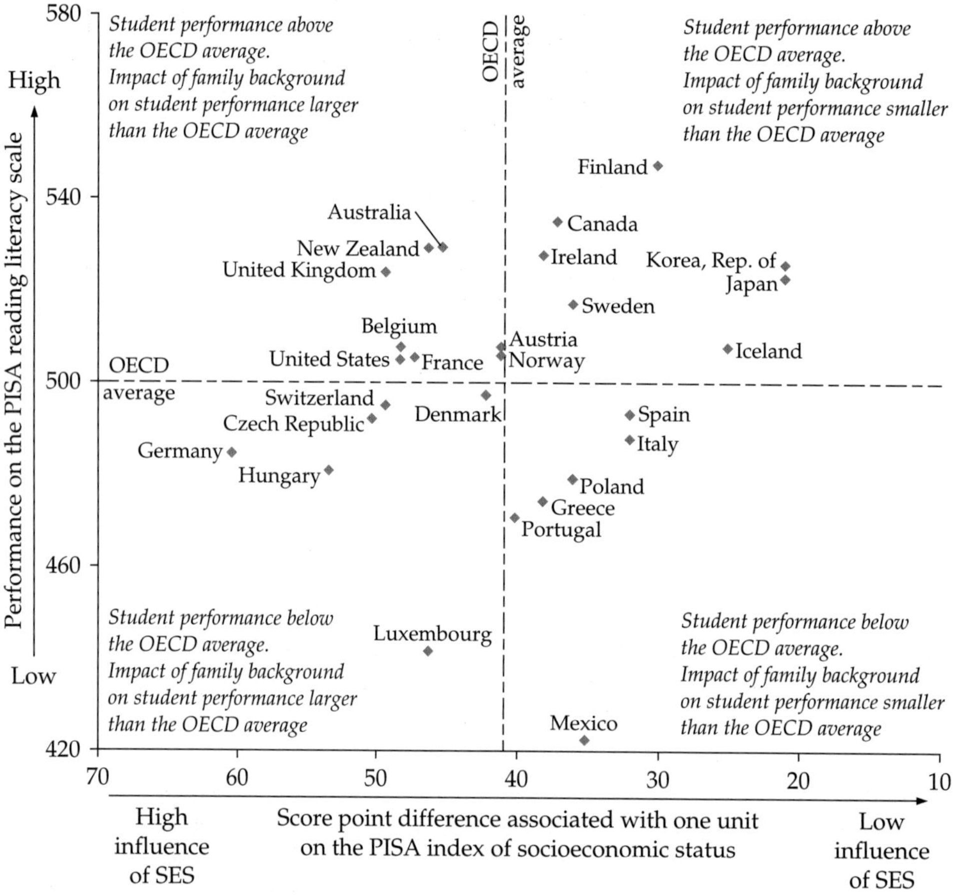

Source: OECD Program for International Student Assessment.

often, however, this is not the case. Survey evidence indicates that 61 percent of temporary U.S. residents who received a science or engineering doctorate in 1998 stayed in the United States during the following five years (1999–2003) (Finn 2005). There are, however, striking differences among countries of origin.[3] For sending countries, benefits depend mostly on the success of national policies in enticing skilled emigrants to return with the new technological and entrepreneurial competencies they have acquired abroad, as well as any newly gained capital and access to international science and technology networks. Also important are national policies for maintaining contact with expatriate workers (box 7.11).

However, diaspora communities may generate significant benefits even if emigrants do not return home. Expatriates do not need to be investors to make an impact in their home country. They can serve as "bridges" by providing access to

3. Stay rates are highest in China (90 percent), India (86 percent), and Eastern Europe and Argentina (69 percent each). Countries such as Brazil, Mexico, Taiwan (China), and the Republic of Korea have significantly lower stay rates.

Box 7.11 Initiatives to Encourage Highly Skilled Migrants to Return Home

Some newly industrialized countries such as Singapore, Taiwan (China), and the Republic of Korea have been successful in encouraging emigrants to return by opening up their economies and implementing policies to foster domestic investment in innovation and research and development. The Taiwanese government, for example, has played an important role in attracting students to return, especially American-trained scientists and engineers, who in turn have helped develop the country's ICT sector. The government set up the National Youth Commission, an information clearinghouse for potential employers and for returning scholars seeking employment. An airfare subsidy is granted to graduating students and their spouses, plus up to two children, if the students decide to return to Taiwan (China) to develop their career. The commission has also established channels of communication with overseas scholars to simplify recruitment when the need arises. It seems such efforts are successful, with an increase in the share of scientists and engineers with doctorates from American institutions who are returning to their country.

Source: OECD 2002; Aggarwal and others 2004.

markets, attracting sources of investment, and sharing their expertise. Influential members of the diaspora can also shape public debate, articulate reform plans, and help implement reforms and new projects.

Policy expertise and managerial and marketing knowledge are the most significant resources of diaspora networks. Such networks' focus on knowledge and policy contributions is a distinguishing feature of a new generation of programs and projects that aim to transform "brain drain" into "brain gain." For example, in a process known as "brain circulation," expatriates—aided by the lowered transaction costs associated with digitization—are transferring technical and institutional know-how between distant regional economies faster and more flexibly than most large corporations. As a positive example of this phenomenon, Chinese, Taiwanese, and Indian engineers are accelerating the development of the IT industries in their home countries. Diaspora networks and expatriates can also be mobilized in other ways, as exemplified by the Global Scot network (box 7.12).

Forms of international trade in higher education such as e-learning programs and satellite schools preclude the physical relocation of emigrant students to developed countries. This lowers the risk of brain drain and offers a greater chance that students' countries of origin will benefit from international educational services and learning infrastructures. E-learning programs are generally supplemented by face-to-face teaching in local tertiary or vocational education institutions. International knowledge transfer at the institutional level typically involves a tertiary education institution (the most active players are in Australia, the United Kingdom, and the United States) that establishes a commercial presence (by opening a satellite school or campus or acquiring all or part of a foreign educational institution) in a host country. Such projects are often carried out in partnership with the local tertiary education institutions.

Box 7.12 The Global Scot Network

Global Scot, part of the Scottish economic development agency, is a highly innovative and successful network of about 850 high-powered Scots throughout the world who use their expertise and influence to help generate new projects in Scotland. The following examples show the role of diaspora members:

- An investment project identified by one of the first members to join Global Scot brought an Internet licensing company to Glasgow. The company, which initially employed eight people, will "quickly become a multimillion pound business," according to its founder.
- At a crucial stage of its negotiations with a U.S. blue chip company, a Scottish electronic engineering company received, within a day of its request, a full day of advice on how to negotiate a licensing deal.
- A specialist provider of training for the international oil and gas industry looking for an entry point into the Gulf of Mexico was connected to a Global Scot member who was the former president of Enterprise Oil, Gulf of Mexico. The member introduced the training provider to oil and gas companies in the region, leading to business associations with several of the companies and a firm foothold in the market.
- A company specializing in creating virtual characters for gaming software made valuable connections with a number of Global Scot members during a trip to California for an exhibition. A director at the company described the contacts as "an absolute bull's-eye target for the type of business advice needed," providing access to "people you would never dream of trying to reach, as there would usually be about a dozen gatekeepers between you."

Source: McRae 2006.

8

Strategic Directions for Selected Countries and Country Groups

This chapter provides concrete examples of knowledge and innovation strategies to be developed across four broad country groups and covering a large variety of situations seen in the developing world.

The chapter begins with the two emerging giants: China and India. These nations deserve close analysis both because of their size and importance to the world economy, and because their experience provides lessons of a more general nature for the developing world. Both illustrate how an enduring development process can start from localized development, catalyzed by mutually reinforcing, incremental economic reforms and knowledge economy (KE)–related investments. Further reforms and KE-related efforts are now essential if China and India are to sustain their position as major world powers. They must face the challenge of growing inequality and be adept enough to field policies of great variety, ranging from those necessary to beat the top global competitors to those required to feed the very poor at home. How China and India cope with such issues and challenges will differ, owing to their very different societal settings.

Two exemplary groups of middle-income countries are then discussed. The first is composed of countries that have been active KE-style reformers in the past, including Chile, Malaysia, Tunisia, and some of the countries of Central Europe that are now part of the European Union. Thanks to their efforts, most have enjoyed significant growth rates. They now need to launch new, knowledge- and innovation-centered reforms if they wish to maintain their growth momentum. To do this, they need to adopt worldwide best practices, and, in some cases, revise the role that their governments have played thus far. The second group discussed includes large economies with a strong science and technology potential, such as the Russian Federation and the Latin American powerhouses—Brazil, Mexico, and Argentina—that have, by and large, been unable to exploit their true KE potential because of inappropriate economic and institutional frameworks. They require well-focused reforms and initiatives based on all four of the KE pillars discussed in chapter 2—uprooting entrenched interests—in order to realize their unexploited growth potential.

Finally, the focus turns to low-income countries, particularly those in Sub-Saharan Africa. African economies have enjoyed relatively high growth rates in recent years, but their long-term development requires articulated, KE-inspired strategies adapted to nations' individual knowledge resources and institutional capabilities. These must be based on both macro-level reforms and initiatives supporting micro-level development. Broad reforms must promote (a) improvements in the business and institutional environment, (b) investment in information and communication technology (ICT) infrastructure, and (c) efforts to reduce illiteracy while increasing higher education capabilities. There is also a need to invest in technological improvements that satisfy basic needs (health, nutrition, and so on) while making the best use of comparative advantages in agriculture, tourism, culture, and ICT (such as call centers).

Regardless of country grouping, to get action on a KE agenda, countries must implement both top-down initiatives based on the four KE pillars and bottom-up initiatives to strengthen and scale up worthwhile local action.

China and India: The Emerging Giants

The emergence of China and India as global competitors is a major development story of our times. By taking advantage of globalization and opening up to the world economy, these economies have—in a few decades—gone a long way toward recovering their historical prominence in the world (figure 8.1).

Several differences between the two nations are worth noting. China opened up its economy and embarked on its ambitious reform path in 1978, while India's major reforms started in 1991. Each has specialized in a different economic sector—manufacturing in China and services in India. Growth was a result of having a critical mass of highly qualified, technologically savvy personnel in India and low- to medium-skilled workers in China. China's growth has been built on a mainly export-oriented economic model, strongly marked by openness to foreign trade and investment, whereas India has been comparatively less open. Finally, the greatest difference is the overall political context: business-oriented socialism in China, and a bureaucratic democracy in India.

Figure 8.1 *GDP Trends of Major Economic Powers over Two Millennia*

Share of global GDP (1990 international dollars)

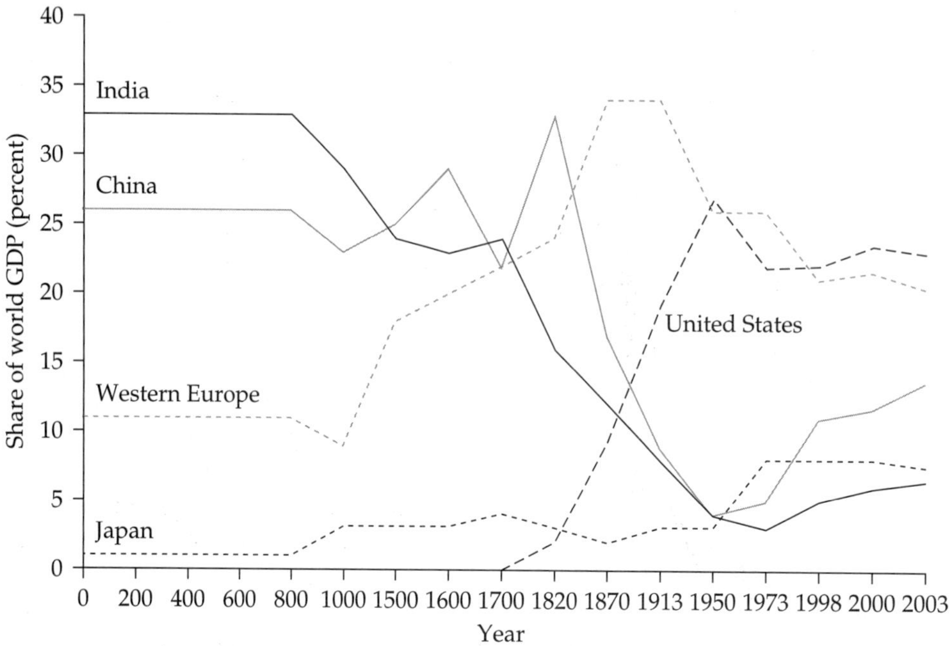

Source: Maddison 2006.

There are also many similarities in the two countries' growth process. Both show the importance of local-level reforms; in both cases, the local level is where things changed first. This is important, because reform is easier to implement on a small scale. When scaled up to the national level after being proven successful at the local, even small efforts can make a big difference. Indeed, the growth and reform process was first implemented in special economic zones (SEZs) in China, and in the technology hub of Bangalore in India. What is interesting to note is that these locations did not grow randomly. The first four SEZs in China and the city of Bangalore in India are strongly connected to the Chinese and Indian diasporas—who, with different roles in each country, were critical in both.

China and India also count on different strengths: India has a strong base of innovative capability, a large pool of large global and national enterprises, a vibrant democratic base, a large English-speaking population, a rapidly mounting middle class with a relatively well developed domestic market, and a relatively young population. China's strengths lie more with its government's strong pragmatism, flexibility, and leadership, as well as its ability to successfully woo the global business world by attracting foreign capital and investment. The countries also face some important challenges in the near future. Securing reliable and sustainable energy sources to fuel their high development rates is a challenge for both, as is translating growth into significant poverty reduction and well-being for the entire population.

China

Although the Chinese government has made significant progress in many areas, several daunting challenges remain, including specific KE challenges. These can only be successfully tackled if there is a further repositioning of the government within the economy and society as a whole. In short, the government needs to take the position of captain, and leave the rowing to others.

Institutional, economic, and regulatory improvements. China's incentives and institutions, despite significant progress over the reform period, still constrain the economy, preventing it from taking full advantage of rapid advances made on the three other KE pillars. Two elements are key: the legal framework lacks clarity, and the institutions needed to accompany a socialist market economy are lacking or underdeveloped. Although legislation is often in place—as for example, the relatively complete set of laws and regulations covering economic rights and intellectual property rights (IPRs)—its enforcement is still inadequate. Appropriate regulations are needed to deal with safety, standards, and environmental issues. Those regulations must be enforced. Market-supporting institutions, such as fair trade and antimonopoly commissions, also need to be strengthened.

ICT policies—broader diffusion to benefit all. Although China's overall ICT penetration is growing rapidly, rural–urban and regional gaps are huge. While the prosperous east coast and second-tier inland cities boast telephone penetration rates of 40–70 percent—well above the national average of 20 percent—some 15 percent of villages, home to at least 120 million people, do not even have access to a public phone. To better facilitate the rapidly developing ICT market, China should streamline competing regulatory bodies, increase the transparency of the decision-making process, and clarify agencies' responsibilities. The Ministry of Information

Industries (MII) could possibly be rolled up into a state communications commission along the lines of the U.S. Federal Communications Commission. This could help to achieve convergence of the telecommunications and cable industries, expand access for rural communities, and promote greater ICT use.

MOVING UP THE VALUE CHAIN—CREATING A MORE EFFICIENT INNOVATION SYSTEM. For the most part, China is still an imitator or adaptor of foreign technologies, with most innovation occurring in processes rather than in products. Aware of this weakness, the government intends to boost the national R&D effort to 2 percent of GDP within the next few years. China's existing research and development (R&D) efforts are in many ways not relevant to industrial needs. For example, out of 100 science and technology (S&T) Ph.D. dissertations in the United States, at least one can be commercialized as a new product. In China, however, this ratio is only one in 500 (Saxenian 2005). R&D efficiency is also low: despite substantial investment in the innovation process, progress is slow on the output side, at least as measured in patents. In addition, technologies in China's elite R&D labs and high-technology parks are not being effectively disseminated or commercialized.

To transform itself from a manufacturing to an innovation superpower, China should implement a clear, publicly funded research strategy. It should strengthen the demand-orientation of existing public research institutes and universities (through more joint research, collaborations, and personnel exchange between research organizations and the business sector, including foreign corporations) and strengthen financial support for innovation. It should also increase R&D funding aimed at small-scale enterprises; liberate the financing constraints inhibiting small and medium enterprises (SMEs), for example, by widening access to stock markets; and establish mechanisms to facilitate technology adaptation and dissemination projects in defined geographic areas before scaling them up nationwide if proven successful. Moreover, access to advanced foreign technology could be improved through efficient arrangements with foreign R&D centers established in China, as well as increased involvement of the scientific diaspora.

EDUCATION AND TRAINING POLICIES. China needs to reduce inequalities in access to education. The adult illiteracy rate ranges from below 10 percent for most of the eastern region to more than 20 percent in many western provinces. The education level of the labor force also varies widely across regions (table 8.1). Education spending per student is unequal; for example, spending per student in primary schools across provinces was 10 times more in Shanghai than in Henan province in 2003 (Dahlman, Zeng, and Wang 2007). To solve this, the government could

Table 8.1 *Education Levels of Chinese Labor by Region, 2004 (percent)*

Regions	*Illiterate*	*Primary*	*Junior secondary*	*Senior secondary*	*College or above*
East	4.1	20.8	45.6	18.0	11.6
Middle	4.5	25.7	49.3	13.9	6.6
West	14.2	35.7	33.2	10.2	6.7
National average	*6.2*	*27.4*	*45.8*	*13.4*	*7.2*

Source: Authors' calculations based on data from China Labor Statistical Yearbook 2005, National Bureau of Statistics, and Ministry of Labor and Social Security of China, China Statistics Press, 2003.

increase funding while improving transparency and accountability related to its use. In addition, compulsory education should be extended from 9 years to 12.

China is boosting its university systems by focusing its support on 100 "elite" establishments. It also needs to significantly improve the quality of its tertiary education system by reforming curricula, pedagogy, and governance. More emphasis should be put on fostering students' creative and innovative thinking and problem-solving skills, as well as their teamwork and communication abilities. Finally, there is a need to develop an effective lifelong learning system, tapping the potential of private provision and distance education and encouraging multiple providers and multiple pathways to education. All of this will require better coordination among local governments and various stakeholders; better assessments, accreditation, certification, and vocational qualification systems; and better information services (box 8.1) (Dahlman, Zeng, and Wang 2007).

Box 8.1 Lifelong Learning in China: The Need for a Broad Action Plan

China needs an effective system of lifelong learning for two main reasons. One is the greater global emphasis on knowledge, which makes lifelong learning essential for enhancing China's global competitiveness. The other is the tremendous range of challenges facing education and training in this new context. Those challenges include:

Competitive pressure. With a trade share of 70 percent of GDP, China is subject to the tremendous competitive pressure of the global system.

Large size. Included in its population of 1.3 billion, China has a school-age population of nearly 260 million—from preschool to tertiary students—and a labor force of 750 million that requires skills upgrading.

Low educational attainment. The average educational attainment in China is still low—at seven years compared with 12 among the countries of the Organisation for Economic Co-operation and Development. This difference is particularly pronounced among the working-age population.

Large regional disparities in income and educational levels. China is very diverse, with richer coastal provinces and poorer western provinces. The education gap between these areas is wide and growing.

Transition to a market economy. With China's transition from a centrally planned to a market economy, all kinds of new market-related skills have to be imparted—not only to students, but also to workers in the labor force.

Massive restructuring and rising unemployment. As the result of a dramatic structural transformation from agriculture to industry and services, millions of rural migrants and workers—laid off from state-owned enterprises—have to be trained for jobs.

Fragmented system of education and training. Like most countries, China has a fragmented system of education and training, with many different ministries and poor coordination across them, leading to waste and duplication.

Financial constraints limit government's ability to address lifelong learning. The financial needs of education and training in China are beyond the means of the government, and many local governments have a problem financing even compulsory education.

To cope with these challenges, great efforts have been made to build a lifelong learning system in China, with significant progress particularly in the areas of private education, distance learning, and large-scale retraining. More systematic reforms and an integrated approach are necessary however. To develop an effective system of lifelong learning, China needs a new role for government—from the main provider of education and training to the overall architect and facilitator of a more complex system of multiple pathways and providers. The government needs to build close partnerships with the private sector and other stakeholders in order to:

(continued on next page)

Box 8.1 (continued)

Set up the rules of the game, ensuring quality, relevance, efficiency, and equity in education and training. Systems of accreditation, certification, and standards, and a comprehensive qualification framework should be in place to ensure quality and relevance. The increasing regional and rural–urban gaps have to be addressed.

Provide accurate and timely information on changing market demands; employment opportunities; and the quality, performance, and offerings of education and learning providers (including international ones).

Provide sufficient funding to meet the increasing demand for education and training. While the government should focus more on compulsory education, the private sector can play a bigger role in higher education and training.

Harness the potential of distance education. Distance education is expanding rapidly, but more attention needs to be given to improving the quality and recognition of distance education in the society and labor market.

To implement these measures, several steps need to be taken, notably setting up systems and institutions that self-adjust to rapid change; defining the responsibilities of key stakeholders; providing appropriate governance, guided by a sound policy framework; integrating into the evolving global system; increasing awareness of what a lifelong learning system is and why and how to build one; getting greater stakeholder buy-in through a nationwide consultation process; improving the instruments—especially the regulations—of the national qualification framework; and creating stronger links with the labor, information services, and financing markets.

Source: Dahlman, Zeng, and Wang 2007.

India

India has recently experienced impressive growth—above 7.5 percent annually during 2002–05. The challenge ahead will be to broaden the base of those who benefit from such rapid growth. India is a dual economy: on the one hand it is a nuclear and space power and a leading developer and provider of ICT services; on the other, it is plagued by illiteracy rates of 52 percent among women and 27 percent among men, and 35 percent of its population live below the international poverty line of $1 per day. Figure 8.2 presents a simple representation of the Indian economy. Only 8 percent of the labor force of roughly 430 million is in the so-called organized sector, leaving 92 percent in the unorganized sector, the bulk in low-productivity subsistence agriculture and services.[1] In addition, over 53 percent of the workforce is self-employed.

STILL TOO CLOSED TO TRADE AND FOREIGN DIRECT INVESTMENT. Despite India's recent economic growth, it is a relatively closed economy. Trade as a percentage of GDP is not high (about 33 percent, compared with 79 percent in China in 2004). Its exports of goods and services—at about 15 percent of GDP in 2004—are less than half

1. The organized sector includes enterprises with more than 10 employees, subject to various forms of government regulation and encompassing modern agriculture, business and services, and government. In fact, about two-thirds of employment in the organized sector is in government services.

Figure 8.2 Stylized Structure of the Indian Economy

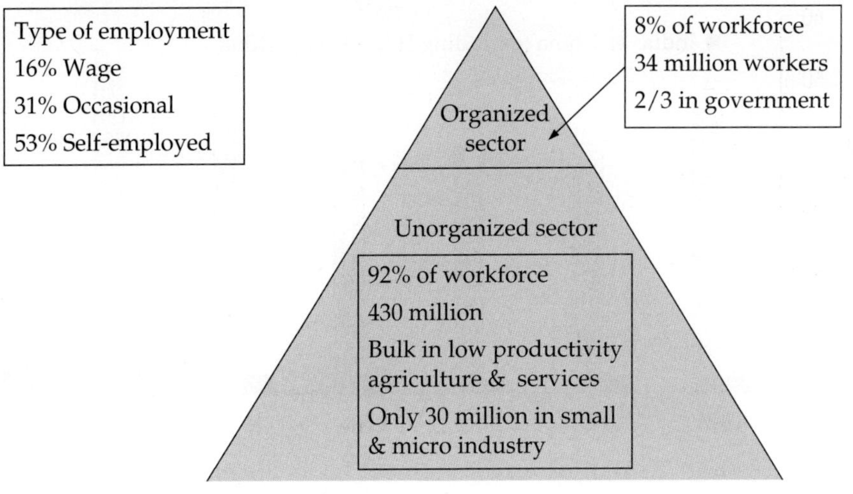

Source: Dutz 2007.

of those of China (40 percent). Although this means that India is somewhat protected from global trends, the downside is that it benefits less from strong foreign competitive pressures to improve performance, and from cost-effective foreign inputs such as capital goods, components, products, and foreign investment—which embody more advanced knowledge. As a result, India is losing market share to major competitors, especially China. In terms of foreign direct investment (FDI), India has a long way to go to catch up with China (figure 8.3).[2]

POOR INFRASTRUCTURE. India's poor infrastructure has often been cited as a major hindrance to economic growth. For example, the transportation infrastructure (roads, ports, airports) does not meet that of China. There is no "cold chain" for the transportation of agricultural products, which spoil on slow-moving highways. This has stymied the development of a national food-processing industry and contributed to the wasting of 25 percent of India's annual agricultural production. Access to reliable power at a reasonable cost is another prime concern for most Indian businesses. But there has been some progress. India's Golden Quadrilateral Highway connects Delhi with Mumbai, Chennai, and Kolkata, and marks the beginning of a series of national road projects (Elliott 2005). The importance of strengthening infrastructure for sustained economic development is clearly recognized: the nation's 2006 and 2007 budgets acknowledge the importance of improving India's roads and highways, ports, telecommunications, and power.

2. India imposes caps on FDI in many sectors such as insurance, aviation, coal mining, media, and retailing. The government is considering reviewing the FDI caps in real estate and telecommunications. Its new minerals policy would emphasize the promotion of FDI and modern technology in the mining sector. Paving the way for FDI to flow into branded retail sales, the government has issued guidelines governing such investments. Rules have also been issued to permit FDI in areas such as airports, natural gas/LNG pipelines, and cash-and-carry wholesale trading. It is important that the government should stay the course and open up more sectors to FDI (www.ibef.org).

Figure 8.3 Inward FDI Flows: India and China

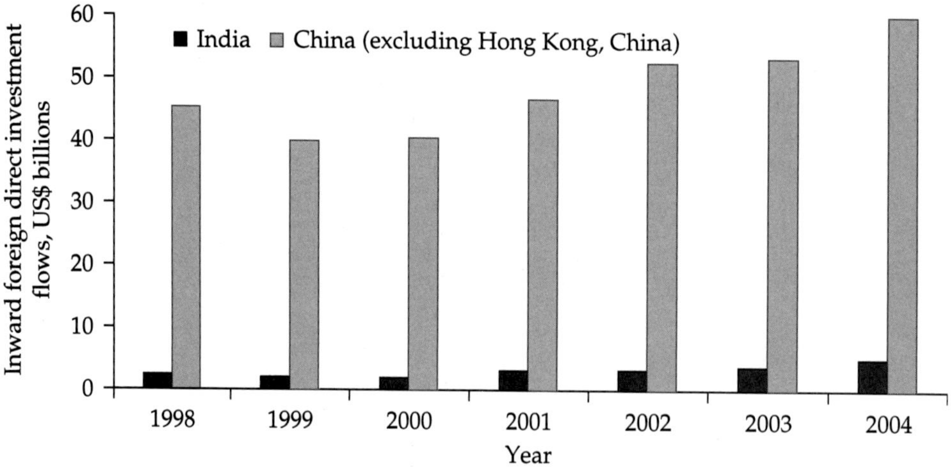

Source: Economist 2005.

Bureaucracy and regulations. In order to capitalize on opportunities, India needs to continue to foster a good investment climate that encourages firms to invest by removing unjustified costs, risks, and barriers to competition. According to *Doing Business in 2006* (World Bank 2005a), Indian entrepreneurs can expect to go through 11 steps to launch a business, a process requiring more than 71 days on average, at a cost equal to 61.7 percent of gross national income per capita. In China, entrepreneurs can expect 13 steps to launch a business over 48 days on average, at a cost equal to 13.6 percent of GNI per capita. India also has one of the more regulated labor markets in the world—rigid labor laws and restrictions on the hiring and firing of workers greatly impede doing business. One key issue is a rule preventing any company with more than 100 employees from discharging workers without obtaining approval from local labor boards, ostensibly to protect workers from unscrupulous employers. The restriction makes employers wary of taking on new staff, opening new factories, or, in the case of smaller companies, growing beyond the threshold of 100. But some progress is being made (Dahlman and Utz 2005).

Education in demand at all levels. India possesses a large pool of highly educated and vocationally qualified people who are making their mark in science, engineering, information technology (IT), and R&D, but they make up a small fraction of the population. To address this, the government initiated a national program for universal elementary education, Sarva Shiksha Abhiyan (Education for All). In higher education, India has a top-quality university system that includes many world-class institutions, such as the Indian Institutes of Technology, Indian Institutes of Management, and the Indian Institute of Science. But India faces an increasing skills deficit. While India (and China) are producing large numbers of engineering graduates, there are serious deficiencies in the training received at the institutions of both nations. Given the great demand for high-quality education, especially in science and engineering studies, and an immense pool of students with high aptitudes, the expansion of high-caliber institutions is warranted so that such institutions can become the hubs of national education chains (Pai 2004). The private sector should also play an increasingly important role in improving

***Table 8.2** Comparative Innovation Performance: India and China*

Variable in Knowledge Assessment Methodology	*India*	*China*
FDI Inflows as % of GDP, 2000–04	0.68	3.89
Royalty and license fee payments (US$/pop.), 2004	0.40	3.47
Royalty and license fee receipts (US$/pop.), 2004	0.02	0.18
Researchers in R&D/mil. people, 2004	119.00	708.00
Total expenditure for R&D as % of GDP, 2004	0.85	1.44
Private sector spending on R&D (1–7 scale), 2006	4.20	3.60
Manufacturing trade as % of GDP, 2004	15.29	50.35
High-tech exports as % of manufacturing exports, 2004	4.90	29.80
Scientific and technical journal articles/mil. people, 2003	12.00	22.65
Patents granted by USPTO/mil. people, avg 2001–05	0.30	0.35
Availability of venture capital (1–7 scale), 2006	4.60	2.70
University-firm research collaboration (1–7 scale), 2006	3.60	3.90
Firm-level technology absorption (1–7 scale), 2006	5.80	5.10
Value chain presence (1–7 scale), 2006	5.10	3.70

Source: KAM version 2007.

Note: The higher the value for the data in the KAM, including qualitative variables, the better a country's performance on that variable. For a discussion of the KAM, see chapter 2.

educational quality. The government should encourage private financing of education and training. It can do this by implementing accreditation systems for private education and training providers and developing an effective system for assessing and certifying vocational qualifications. In a laudable move, the government has recently backed 100 percent FDI in higher education in India ("Government for 100% FDI in Higher Education," *Hindustan Times,* February 6, 2007).

An innovation system in need of reform and expansion. India spends only a small fraction of its GDP on R&D (0.85 percent of GDP in 2002), receives a small sum of worldwide royalty and license fees, and has a low rate of publishing scientific and technical articles in mainstream journals (table 8.2). India's share of global patenting is also small; despite having a strong R&D infrastructure, India is weak in turning its research into profitable applications. In recent years, it has taken steps to strengthen its R&D infrastructure (see chapter 5), develop technological innovations, and alter the mindset of its people toward better creation, acquisition, and use of technology. But it needs to improve its innovation system further by tapping knowledge from abroad through channels such as FDI; technology licensing; the importation of capital goods that embody knowledge, such as advanced products, components, and services; and the dissemination of such goods for greater economic and social development. It should also strengthen university–industry links so that research can focus on the needs of the economy. Box 8.2 highlights the case of Bangalore, which, surprisingly, is not characterized by strong university–industry links except in a few sectors. To move ahead in the global innovation race, India must also ensure rigorous enforcement of IPRs, as this is a key to attracting new investments into the country. To its credit, India is examining its innovation system, and looking at ways to strengthen its innovation environment (Dutz forthcoming).

With hundreds of millions of poor people living in rural and urban areas, there is a need for massive technology-diffusion efforts and the creation of advisory services for business creation, technical design, commercialization, management, and

Box 8.2 Bangalore's Innovation System

Few would doubt that Bangalore's success with information technology (IT) is embedded in its educational infrastructure and research institutions. But the presumed synergy of industry and local institutions is weaker than one might expect. Beyond the supply of technical talent to the IT industry, university–industry links are few and weak, as revealed by a survey of seven major research institutions and universities. Such links were mostly confined to pharmaceutical and chemical industries. Nonroutine, knowledge-intensive endeavors are needed, including the cross-fertilization of new ideas and new modes of institutional interaction among industry, academia, and government.

Bangalore's dynamism stems from linear and extensive—rather than nonlinear and intensive—growth. Its overt export dependence and heavy reliance on the American market is conducive to extensive growth as the local educational infrastructure responds favorably to the demand for technical talent. Indian firms can take on large, high-volume, low-value projects because of the availability of IT workers. Selectively, some firms are establishing limited ties to universities, and many multinational firms are establishing R&D centers in Bangalore. These initiatives, however, are either limited in scope, sporadic, or constrain technological spin-offs in the regional economy because of their enclave nature. Extensive growth would be sustainable only on the unrealistic assumption that the global IT industry will change little, or that the supply of skilled workers and the demand for low-cost services will remain inexhaustible. Skill shortages have already appeared globally—and in India. India is expected to face competition from low-cost producers such as China, Vietnam, Eastern Europe, and Russia. Hence, a strategic shift toward intensive growth, driven by innovative capability, is necessary. With competition, the boundaries between basic and applied research blur, making academia, industry, and government key agents in generating knowledge and technologies. This trilateral institutional architecture, or triple-helix model—theoretically viewed as an engine of innovation and intensive growth—has been largely absent in Bangalore.

Source: D'Costa 2005.

finance. Some initiatives, such as the Honeybee network that protects indigenous knowledge (Gupta 2006), show the way ahead but remain too small in the face of the country's great needs. A large-scale, nationwide program should be launched with municipal powers combining technical support (in mobilizing public laboratories, universities, and so on), and financial assistance (notably through microcredit schemes).

Active Reformers

The middle-income countries that we have chosen to classify as "active reformers" include Chile, Malaysia, Tunisia, Jordan, Mauritius, Costa Rica, and several Eastern European countries. Of the latter, notable are Slovenia, Hungary, Estonia, the Czech Republic, and Slovakia. Among these nations' many differences are size—in terms of both population and territory—specialization, culture, geopolitical context, and the stage reached in the transition toward a knowledge economy.

Of the two strong characteristics that unite them, the first is performance. These countries have done relatively well, and have thus far been able to capitalize on their internal strengths—whether through reforms or historical legacies—and the availability of cheap labor to induce growth. Their second defining factor is openness to

Figure 8.4 *WEF Growth Competitiveness Index Rankings: Active Reformers*

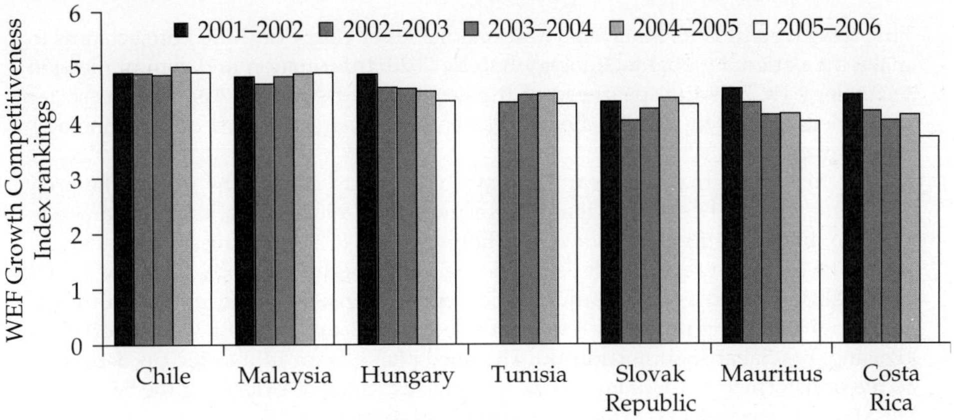

Source: WEF 2006.

external influences. These countries have often capitalized on outside knowledge for their development, mainly through FDI and sometimes through long-lasting links with advanced countries.

Most, however, either did not fully exploit opportunities offered by FDI for true technology and knowledge transfer, or did not invest enough in building domestic innovation capabilities. As a consequence of their specialization in low- and medium-tech manufacturing—and inadequate technology adoption from FDI and domestic R&D and innovation efforts—several active reformers are now experiencing difficulties in maintaining their competitiveness, as shown by the evolution of their ranking on the Growth Competitiveness Index of the World Economic Forum (WEF) (figure 8.4).

A low level of business R&D is a sign of insufficient indigenous innovation capabilities—less than 0.3 percent of GDP in the active reformers compared with more than 1 percent in advanced economies.[3] Their innovative performance, as measured by various outputs (patents, innovative products) is also modest. The major issue, then, is the development and growth of innovative clusters that will allow these countries to sustain their growth momentum. The measures needed to address these issues will differ according to economic specialization and cultural background. A few key examples illustrate different approaches to increasing economic performance.

Developing Countries: Malaysia, Tunisia, and Chile

Both Malaysia and Tunisia—and, until recently, Chile—have elaborated ambitious KE strategies over the medium term.[4] For Malaysia, two main issues could

---

3. Among the latter, Ireland and Singapore have followed a similar strategy of attracting FDI and suffer from the same syndrome. Ireland is trying to overcome the problem by investing massively in R&D, and Singapore has undertaken a major overhaul of its university, making it an entrepreneurial university—an effort that is beginning to bear fruit.

4. In 2005 Malaysia elaborated a KE master plan that included a long list of policy measures covering the different pillars; and Tunisia has introduced KE as a major dimension of its five-year plan.

Box 8.3 Malaysia's Multimedia Super Corridor and Expansion Project

First conceived in 1996, Malaysia's Multimedia Super Corridor (MSC) project aims to make the nation a global technology hub by 2020. Information and communication technology (ICT) is the spearhead of the development strategy. Applications so far developed have been Smart Schools, Telehealth, e-Business, smart card technology, and e-government.

The MSC's original site, near Putrajaya, was set up during the first of the project's three phases (1996–2003). It now hosts more than 900 multinational corporations (MNCs), including foreign-owned and home-grown ICT companies, and generates some $1.6 billion in revenue while providing 22,000 high-value jobs. Phase two, covering 2003–10, focuses on the duplication and expansion of the project—through industrial clustering processes—to span selected sites and create a web of corridors (Penang, EastSabah, South Sarawak). The third phase, from 2010 to 2020, is dedicated to the transformation of Malaysia into a knowledge society, extending the benefits of the MSC and the web of corridors to the national community.

Various incentives induce "technopreneurs" and investors to set up shop in the MSC. A distinctive and comprehensive package of benefits has been instituted, including tax and fiscal incentives, cyberlaws, and infrastructure investments. This, as well as the overall management of the project, is overseen by the Multimedia Development Corporation (MDeC) which also acts as a one-stop agency with a guaranteed 30-day application procedure.

Source: www.msc.com.my/msc/msc.asp.

be prioritized to facilitate the next quantum leap toward a knowledge economy: taking greater advantage of major investments made in the Multimedia Super Corridor (MSC) (box 8.3), and upgrading university and innovation systems. The MSC was conceived as a test bed for new ideas, and an environment to jumpstart innovation in ICT. On the positive side, it has led to notable improvements in Malaysia's general ICT infrastructure (WEF 2005) and significant employment creation.

Success has remained mostly at the infrastructure development level,[5] however, and further efforts are needed to help the MSC evolve into a fully fledged dynamic cluster. There has, thus far, been low synergy between academia and industry, and too few entrepreneurs armed with high-risk venture funds. Evidence also shows that little R&D is undertaken in the MSC because of the lack of high-tech human capital and appropriate innovation strategies (Lall 2002).

The second priority area is academia. The Malaysian government has recently launched a new and highly visible project with the aim of developing top, world-class universities. First-stage evaluations by the World Bank, however, show that a number of questions need to be addressed: How many such top universities can the country afford? How can frameworks be harmonized between the public and private sector (fairly active in the higher education area)? How can inequality be prevented among students if high fees are introduced to sustain the development of an improved university system? How can lower technical school infrastructure be strengthened, and how does this fit in to the larger university plan?

5. Personal communication with Dr. K. J. John, Malaysia.

***Box 8.4** Tunisia: Developing Technopoles*

The Tunisian government has the ambitious plan to establish 10 "technopoles" throughout the country, each focused on specific sectors and building on existing educational and industrial resources. So far, the most concrete achievements have been made in the telecommunications technopole of El Gazala, developed in a Tunis suburb around a nucleus of existing telecommunications schools. Opened in 1999, the technopole counted 31 enterprises in 2004, of which six were multinationals; 620 jobs have been created in the private sector along with jobs for 300 public sector managers and 110 teachers and researchers. About 50 million dinars have been invested in infrastructure, an amount roughly equal to the annual sales of the firms in the technopole, half of which are export sales. Prospects are good, although problems are perceived in the development of research activities. Direct job creation is moderate, but the concept of the technopole is valid as a means of raising innovation capabilities and pursuing territorial development. The government program should be pursued steadily, with a clear understanding of the critical mass necessary for success.

Source: World Bank 2004g.

Tunisia is illustrative of an economy that has developed a strong industrial base in sectors such as textiles (50 percent of manufacturing employment) and electronics by attracting foreign investors. The success of these industries owes much to a pragmatic policy comprising significant improvements to the business environment, strong tax incentives, and investment in labor force qualifications. These industries, however—and therefore the economy—have been experiencing a certain loss of competitiveness. Moreover, despite a relatively high growth rate—5 percent over the last 10 years—Tunisia is facing huge unemployment rates, notably among university graduates (50,000 arrive in the job market each year, and more than 70 percent do not find a job).[6]

To counter these trends, the Tunisian government has recently made progress on the education and ICT pillars. University enrollment increased from 23 percent in 2001 to 32 percent in 2004, while enrollment in short-term technical programs was up from 21 percent in 2001 to 25 percent in 2004. On the ICT front, Tunisia devotes more than 7 percent of its GDP to ICT expenses. Some 30,000 jobs have been created in the sector over the last four years. Telephone penetration is increasing rapidly (thanks to mobile phones), but computer and Internet penetration is still low (with less than 10 percent of the population using the Internet).

Measures to improve the innovation system include the increase of public R&D expenses, doubling in five years to reach some 0.8 percent of GDP; the implementation of a large-scale program, largely financed by the European Union, to upgrade the technical and organizational capabilities of established enterprises (more than 2,300 enterprises have benefited from the program in 10 years); the planning of 10 technopoles, of which the most advanced is El Gazala, in place since 2000, in the telecommunications sector (box 8.4). In terms of employment, results are not negligible, although many jobs are related to the public sector.

6. For a more detailed analysis of the evolution of Tunisia's economy, see World Bank (2004g).

All these measures, however, have been insufficient to boost R&D; private sector R&D efforts remain very low, at 0.13 percent of GDP. New measures are being introduced to stimulate co-contracting and partnership between local producers and foreign firms to stimulate technology and knowledge transfer, and also to build joint R&D centers.

Governance issues remain a concern for both Malaysia and Tunisia as they seek to unleash local entrepreneurial dynamism and encourage innovation. Indicators from the World Bank's Knowledge Assessment Methodology (discussed in chapter 2) clearly show that Malaysia is falling behind in its ranking on the economic and institutional regime, as does the Economic Freedom Index, which ranks Malaysia in 87th position (Mauritius is 71, Tunisia 67, Thailand 60, and Costa Rica 50). Although Tunisia has made significant improvements in the business environment, improving government efficiency, enterprise creation, employment rules, contract enforcement, and so on, the overall business climate is affected by a lack of development, venture capital, and internal competition—all factors that seriously affect the dynamism of the private sector and its renewal (World Bank 2004g). Moreover, there are serious problems with accountability and press freedom; excessive centralization hinders initiatives at local (city and region) levels, in schools and universities, and among other areas key to innovation and knowledge-based growth.

Chile, on the other hand, is exemplary in the way it has set up its business environment. Ranking at the top of the World Bank's Doing Business indicators, Chile boasts an excellent economic and incentive regime, and is ranked 13th in the Economic Freedom Index, just behind Estonia, and in 20th position—one place after the United States—in Transparency International's Corruption Perception Index 2005. It has also pursued good policies in the functional knowledge pillars, making necessary investments (box 3.3).

Chile's formal R&D base and industry–science collaboration nevertheless remain rather weak. Total expenditure on R&D in 2006 stood at just 0.6 percent of GDP, compared with an average of 2.26 percent in the countries of the Organisation for Economic Co-operation and Development (OECD 2007). An in-depth analysis (Lederman and Maloney 2003) of the Chilean case demonstrates that the nation is underinvesting in R&D, and that it is underperforming in terms of scientific and technical outputs.[7] Chile's underperformance in innovation is partly due to lackluster performance in innovation investments, trapping it in a pattern of natural-resource-based innovation.

To overcome this situation, two basic measures need to be considered. First, there is a need for significant adaptation of the science and innovation system. Chile's science base can be strengthened by rationalizing its support structure and improving interactions with industry. This is being partly addressed by the establishment of Millennium Science Centers with World Bank support (Holm-Nielsen

---

7. Lederman and Maloney (2003) estimate that the economic returns to R&D and to licensing for countries of Chile's level of income are high at around 60 percent. Using the estimated return on physical capital and the U.S. long-run return on stocks as high and low cases for the opportunity cost of investment, simple calculations suggest that Chile should be investing two to eight times more in R&D than it did in the 1990s. Similar results emerge for investments in licensing of foreign technologies. Since the average returns on R&D investment are higher in less developed than in high-income countries, the puzzle is why the former invest so little in R&D.

and Agapitova 2002). There also are issues of institutional overlap and coordination that make long-term planning difficult, as well as business–research collaboration issues—both of which could be resolved through the establishment of an efficient agency on the model of the Finnish National Technology Agency, TEKES, to coordinate most innovation support, with a strong focus on university–industry cooperation. The role played by Fundación Chile could be reconsidered within such a configuration.

Second, there is a need to raise the general education level of the population (Benavente, de Mello, and Mulder 2005). The country's competitiveness issues are partly explained by relatively low university enrollment rates and the relatively low quality of secondary education. Several studies, including Tokman (2004), Eyzaguirre and others (2005), and Fuentes and Mies (2005), conclude that the main problem in Chilean education is quality rather than enrollment rates, with the exception of tertiary enrollment.

Eastern European Transition Economies

Since 1995, Eastern European transition economies have significantly improved their ranking on the Knowledge Economy Index (presented in chapter 2), a trend that contrasts with other transition economies and can clearly be attributed in part to their accession to the European Union (figure 8.5).

For some, such as the Slovak Republic, progress has come because of considerable improvements in the economic and institutional regime; for others, it is due to improvements in innovation systems (Czech Republic and Slovenia), or education systems (Latvia, Lithuania, and Poland). Estonia has made progress on all KE

Figure 8.5 *Improvements in the Knowledge Economy Index in Eastern Europe, 1995–2005*

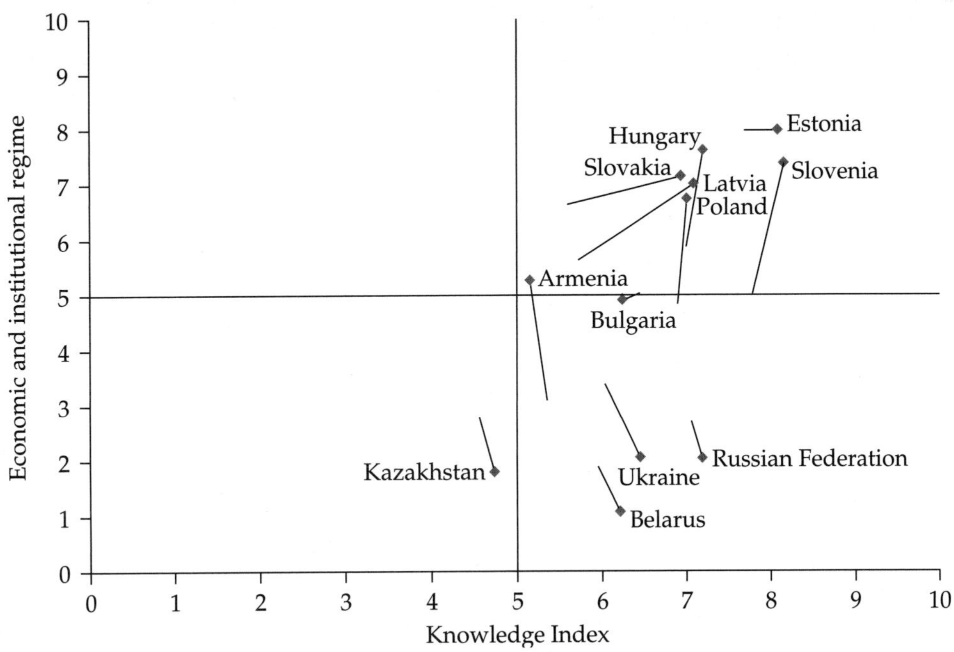

Source: KAM version 2007.

pillars (figure 8.5). The country boasts a notably high percentage of people with tertiary education (30.4 percent as compared to the EU-15 average of 21.2 percent), and a high investment in ICT (11.5 percent of GDP as compared to the EU-15 average of 6.3).[8] Romania and Bulgaria's recent accession to the European Union should stimulate reforms in these two countries.

As a general rule, however, these countries still have to catch up to their Western European counterparts—particularly with respect to labor productivity, which in most cases is less than 40 percent of the EU average. The source of this gap is partly structural—that is, it derives from differences in shares of low- and high-tech sectors—but evidence suggests that the majority of the productivity gap comes from differences in technology, management, and organization rather than industrial structures (European Commission 2003). A cause of further concern, statistics from the European Commission's European Innovation Scoreboard show a growing gap in KE capabilities between new accession states and the EU-15. The chief cause of the widening gap seems to be low levels of investment in public and private R&D, and divergences in the supply of workers with tertiary education (European Commission 2004). Innovation performances are also below the EU average.

A major problem has been a lack of R&D system reform. The structures inherited from the Soviet period (figure 8.7) were fundamentally ill-adapted, but practically no country (except Estonia in the early 1990s) made the needed changes. Academies of science have continued to benefit from privileged support, while university research has remained at a low level of volume and quality. Branch institutes have been left without resources, with no mechanisms to ensure that competences accumulated over years were put to productive use. In fact, MNCs were the main source of support for applied R&D, and in some countries, such as Hungary, where skills were particularly strong, important research collaboration developed for the benefit of Western industry. As a whole, funding of public R&D in Eastern European countries is too low (0.2–0.5 percent compared with the EU average of 0.67 percent).[9]

Second, there were heavy ICT investments during the transition years in many of these countries, notably Estonia, the Czech Republic, Hungary, Poland, and Slovakia, but they have yet to bear fruit.

Finally, basic education systems continue to suffer from training methods that rely too heavily on narrow technical knowledge and rote learning. A recent international survey undertaken by the OECD's Program for International Student Assessments (PISA) highlights the insufficient performance of most transition economies. With the exception of the Czech Republic, all EU transition economies perform in the lower quartile of OECD countries in PISA tests of reading, science, and problem solving (OECD 2004c).

Economies with Untapped Potential

Countries such as Argentina, Brazil, and Mexico, in Latin America, the Russian Federation, and Ukraine, in Eastern Europe, have relatively developed human capital and scientific capabilities but nevertheless have turned in volatile and inconsistent

---

8. The term EU-15 refers to those countries that joined the European Union prior to 2004. This paragraph draws on data and analysis from the European Commission (2005b).
9. See www.trendchart.cordis.lu.

Figure 8.6 Economies with Untapped Potential as Revealed by the Knowledge Economy Index

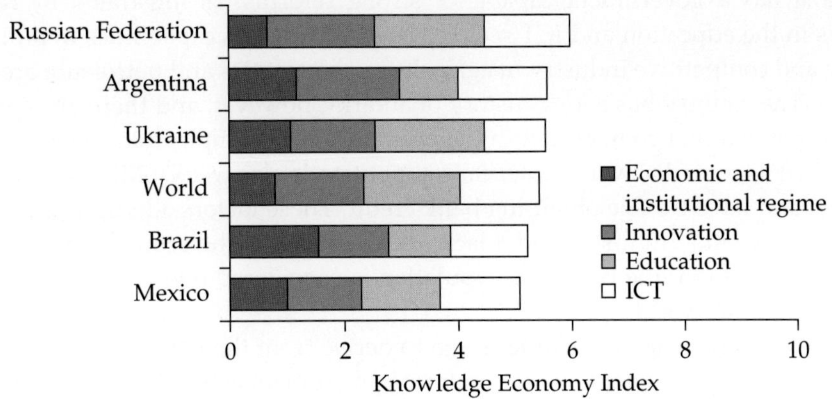

Source: KAM version 2007.

economic performances. Serious weaknesses in their economic and institutional regime (figure 8.6) hinder their capacity to transform their knowledge capabilities into wealth. In this respect, they differ from the regional role models ("active reformers") discussed in the previous section.

Latin American Powerhouses

The three large Latin American economies—Argentina, Brazil, and Mexico—have some common features. In the last 15 years or so, all experienced a severe financial crisis caused by the lack of a solid and in-depth financial base, and those experiences have frustrated long-term investment. All three have high levels of inequality, despite some recent improvement in Brazil. From a KE viewpoint, all have relatively large scientific and engineering communities, of which the S&T performances are not negligible. All have been able to significantly increase their basic education efforts and literacy rates. The difficulties encountered in exploiting their potential for innovation

Figure 8.7 Economies with Untapped Potential as Revealed by the Knowledge Index

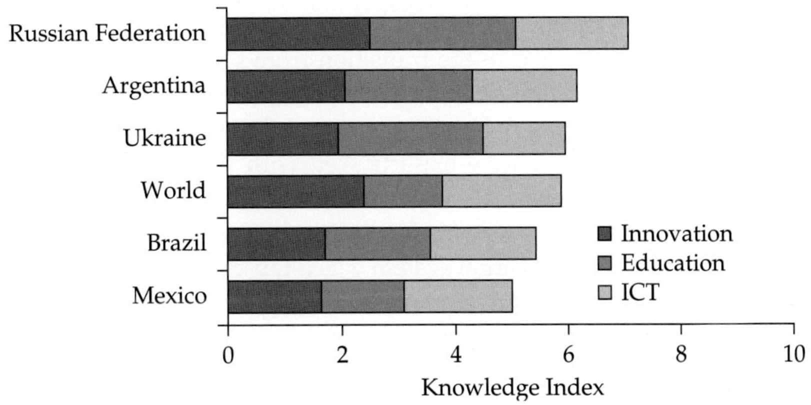

Source: KAM version 2007.

and knowledge-generation appear to come from persistent weaknesses in their economic and institutional regimes.

Brazil has a government capable of strong reforms, as illustrated by recent actions in the education and ICT sector. The government's capabilities in building strong and competitive industry in agriculture, aeronautics and petroleum are also proven. The country has a long history of autarky, however, and there are signs of insufficient external competition in the economy. For example, trade as a percentage of GDP is remarkably low. Internal competition is also weak. SMEs receive little support and have trouble obtaining bank credit. These factors add up to a paucity of innovative small industry and a lack of dynamism in the economy as a whole, despite the fact that the Brazilian population is considered one of the most entrepreneurial in the world.

In Mexico, the major difficulties seem to derive from the inability of the central government to implement necessary reforms. Incumbent enterprises in state-dominated sectors such as telecommunications retain control of markets. Universities (students and professors) are resistant to change. Few actions have been implemented to boost the innovation capabilities of the country, despite the existence of a relatively powerful and independent agency, CONACYT.

Argentina has recovered well from its financial and monetary crisis, but this has little to do with KE factors. While rebuilding from the crisis, the government has not been able to act consistently in KE-related policy domains—industry, education, science, and technology. Most innovative initiatives have come instead from the business community and local authorities. The country's prolonged crisis provoked an exodus of highly educated people, so that Argentina now has strong knowledge diasporas: self-organized communities of expatriate professionals (managers and owners of firms, engineers, and scientists) who wish to assist their home country or region in technology, innovation-based entrepreneurship, science, education, or infrastructure (Kuznetsov 2006).

All three economies have significant pockets of excellence that reflect the potential of the economy at large. The Monterrey urban cluster in Mexico (box 8.5) is a good illustration of geographically concentrated excellence, whereas Petrobras in Brazil is a good example of sector-centered dynamism.

The Countries of the Former Soviet Union

The disparity between knowledge potential and use is particularly evident in the Russian Federation. The former USSR was a leader in S&T, but because its innovation system was not reformed in the transition from socialism, many of its best people emigrated or moved into financial services. The latter could be considered a positive development, if those who migrated from S&T were employed by firms that generated knowledge-intensive exports. Yet, with the exception of military exports, for which the main destinations are India and China, knowledge-intensive exports are insignificant.

Historically, the Soviet innovation system was a top-down affair, mirroring the planning system of the economy as a whole and completely inadapted to a market economy. Knowledge was supposed to flow down from the basic research pursued in the institutes of the Soviet academy of sciences—no university-based research

Box 8.5 *Mexico's Monterrey Urban Cluster*

The Mexican city of Monterrey is located in a barren area with scarce natural resources and an extreme climate. It is the birthplace of Mexican industry and the center of a cluster of industries in metal working, machinery and equipment, chemicals, and ceramics. Many multinational corporations are present, but a significant share of production is accounted for by large domestic conglomerates—among them FEMSA (beverages and retailing), AXA (chemicals, metal, auto parts, food), Proeza (food and automotive), Vitro (glass), Cydsa (chemicals, textiles), Pulsar (biotechnology, financial), Alfa (chemicals, food, auto parts), Imsa (steel, batteries), and CEMEX (cement).

Some of these industrial groups have come together to acquire two of the largest national banks (Bancomer and Banorte), making the city increasingly important as a financial center. Nineteen institutions of higher education train students for careers with employers in Monterrey and around the world. The largest is the Instituto Tecnológico de Estudios Superiores de Monterrey (Tec). Tec was created in 1943 by Alfa, Vitro, and other leading industrial groups to prepare first-rate engineers for their firms. One of the country's leading institutions, TEC maintains close contacts with industry. A key function has been to provide a forum in which representatives of local firms and government agencies can discuss the region's future with academics. This active forum has given Monterrey's leaders a shared vision. All economic agents are currently working to make Monterrey a center for high-technology manufacturing and services.

Source: Dahlman and Kuznetsov 2007.

was allowed. Basic research was supposed to be adapted to applications by branch institutes and design bureaus affiliated with line ministries and finally put to use by enterprises (figure 8.8). The system was organized by sector, with different fields of specialization distributed by design throughout the USSR and Eastern Europe. More than 60 science cities focused on defense-related R&D made up the bulk of the Soviet effort (OECD 1994).

Figure 8.8 *Hierarchical Design and Sector Separation in the Soviet Innovation System*

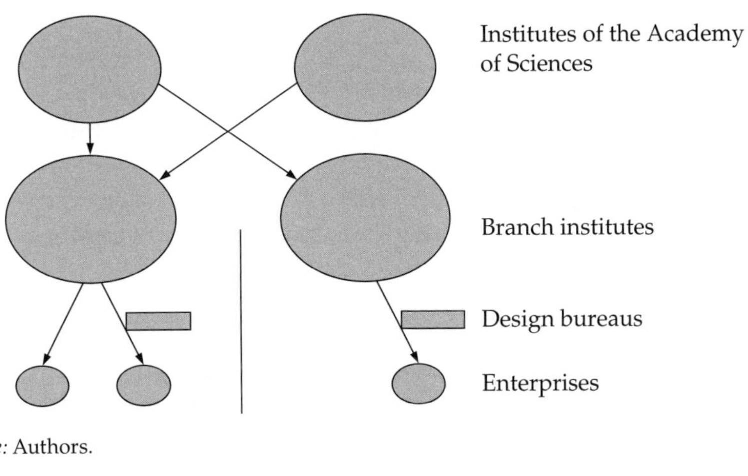

Source: Authors.

Some innovative developments have built on competencies and infrastructure remaining in the science cities. For instance, in Tomsk, an old university city in Russian Siberia with a population of about half a million people, an innovation cluster is developing around small firms specializing in applied mathematics and software development, biotechnology, and advanced materials. Most of the firms are spin-offs of local universities and founded by technical talent with ambition and an entrepreneurial streak. The cluster has developed by bootstrapping—through trial-and-error entrepreneurs have acquired managerial and marketing skills to penetrate demanding international markets. Although some venture capital is available, self-financing remains the rule.

As with the Latin America powerhouses, the main constraint faced by the countries of the former Soviet Union is the economic and institutional regime. Some priority reforms to unleash the untapped KE potential of these economies are outlined below.

Tapping Countries' Knowledge Potential

Changing the rules of the game: credible economic and institutional reforms. A major and concerted effort of reforms and institution-building is needed to unlock the KE potential of the countries discussed above. Reforms to improve the investment climate and allow economic actors to make long-term plans are the first order of business in Argentina and the Russian Federation. Building a better investment climate is a monumental task that will take many years to complete. But changes can bring quick results. Institutional reforms in key areas can become entry points that benefit stakeholders and allow reformers to leverage further support and momentum. Drastic reductions in the costs of establishing, operating, and winding up a business are one such entry point. Both Argentina and Russia have made some progress in deregulation, but more dramatic progress is required. Innovative steps to curb the power of vested interests in government (which thrive on the rents that existing regulations provide) will be required. Success is possible, even against long odds. One example is Mexico, which made deregulation a hallmark of its reform strategy in the late 1980s.

The economies with untapped KE potential also require major reforms in areas as diverse as ICT systems, energy, labor, and finance. The reforms must create an even playing field for competition, ensure the efficient entry and exit of service providers, and provide strong regulatory capacity to secure compliance with minimum standards. ICT infrastructure reforms, for example, can help to reduce the dominance of incumbent providers, such as Telmex in Mexico, through the creation of a more open market and incentives for new entrants. Labor market reforms would ease employment-protection provisions, establish revenue-support systems to cushion workers who lose their jobs, and modernize the collective-bargaining framework.

Releasing innovative capabilities. A general policy priority is to recombine the assets found in existing programs and policy-making institutions. Because industrial assets, ample human capital, and public programs already exist (for SMEs, R&D, linkage promotion, and so on), investment in new assets and programs is less important than recombining viable existing assets in ways that respond more nimbly to current challenges. Doing so could lead to a new generation of programs that draw on existing resources to put grossly underperforming human capital, R&D, and industrial assets to work in sustainable ways.

So-called bridge institutions—the venture capital industry, enterprise incubators, diaspora networks of the highly skilled—that are able to link academic science with industry and local strengths with global opportunities are crucial catalysts in most successful recombinations of existing assets.

Recombined programs often require a few years of institutional learning before they reach their potential. In the Russian Federation, as in many countries, public technology and research institutes are not concerned with industry needs, while the captains of industry are more accustomed to rent-seeking than innovation. Playing by the existing rules provides more payoff than experimenting with technology. Fortunately, however, exceptions are more and more common. On the research side, some applications of R&D are almost ready for adoption by customers in the productive sector, where some industrial groups are interested (if only as a curiosity) in obtaining the quasi-rents available from innovation. It is important to grasp the opportunities that arise, such as the major programs launched by the Russian Ministry of Science and Technology to boost university-industry cooperation in key technologies.

STRENGTHENING HUMAN CAPITAL. Beyond the economic and institutional regime, human capital is the major long-term constraint. Countries such as Brazil and Mexico have made substantial progress in raising educational standards, particularly at the primary and secondary level, but higher education (particularly in technical fields) remains a challenge. What is needed is not greater investment in education but reforms to increase the efficiency of educational spending and improve educational quality. Any successful reform would have to curb the power of teachers' unions and raise the rewards offered for educational quality. The necessary reforms would strengthen regulations and standards pertaining to certification, accreditation, testing, evaluation, and the recognition of prior learning.

Technical education and university–industry links remain a major problem in all of the countries discussed here. Besides teaching and research, universities are increasingly expected to perform a third mission: to contribute directly to economic development and social welfare. The new role requires universities not only to produce but also to commercialize knowledge—that is, to use research results to create intellectual property and contribute to marketable innovations. Many universities already are performing this role—the University of Campinas in Brazil and ITESM in Mexico are just two—but much remains to be done. Here, too, bridge institutions are the key to accomplishing the mission. As efforts to leverage the higher education system are scaled up, the isolated successes can inspire policy measures that link the higher education system to the needs of industry, such as the venture capital industry and the export support industry.

MULTIPLYING AND SCALING-UP LOCAL INITIATIVES AND RELEASING ENTREPRENEURIAL DYNAMISM. In these difficult economic and political circumstances, change comes principally from local and regional initiatives based on learning processes and institutional experimentation. A spectacular example of entrepreneurial self-discovery made possible by government flexibility is provided by the province of Mendoza in Argentina, which has become in a few decades a major producer and exporter of wine of increasingly high quality (box 8.6).

Box 8.6 *Mendoza's "New Industrial Policy"—Public–Private Institutions Providing Public Goods and Services*

In a little more than 10 years, the Argentine wine industry has staged a massive turnaround, transforming itself from a large-volume producer of cheap, poor-quality wine with almost no export potential to a continuously improving industry that commands some 3 percent of the $12 billion international wine market. The transformation is led by the province of Mendoza, which accounts for 90 percent of Argentine wine exports—much greater than its share of Argentina's production. The export boom stems largely from product innovation and vast improvements in quality. More than 70 percent of exports go to the United States, European Union, and Japan—all sophisticated markets. To achieve such gains, Mendoza's firms and farmers have had to implement continuous process and product improvements throughout the value chain, from vine selection and planting to careful vineyard maintenance, flawless harvests, precise fermentation, and judicious blending. The transformation in Mendoza is all the more remarkable given that the neighboring province of San Juan, with the same tradition of wine making and the same favorable climate, has not experienced it. The two provinces are embarked on diverging trajectories.

The first step toward the successful new industrial policy of the Argentine province of Mendoza came in the late 1980s, when the newly elected provincial government decided not to rapidly privatize its large, bankrupt winery. The government's choice to convert the winery into a federation of cooperatives grew out of an earlier campaign to promote the self-organization of thousands of struggling grape growers. The key lesson here was that a lasting solution to a persistent crisis came only after the government invited a range of stakeholder groups into the restructuring process.

As Mendoza confronted other agricultural problems, it solved them through new public–private institutions that were focused on providing a set of public goods and services—R&D, training, export promotion, joint-marketing, and insurance—to help people produce wealth from knowledge. But in order to overcome past animosities and self-dealing, the institutions have had to have three characteristics.

- All are public–private. Their legal form, resources, and governance elicit the participation of government agencies, sectoral associations, and other nongovernmental groups.
- Membership is inclusive. Even associations from backward regions enjoy voice and access.
- In return for membership, participants adhere to rules of deliberation and collective problem solving.

By the late 1990s, the new institutions had helped transform even the most backward parts of Mendoza's viticulture while forging new ties between old institutions, associations, and localities in the province. But the revolution did not stop at the provincial borders. The institutional model pioneered by Mendoza is being replicated at the national level. In 2004, the Argentine congress and president signed into law a self-financing wine sector policy that is governed by a nongovernmental body comprised of representatives from relevant business associations, research institutions, and provincial and federal ministries. This policy and governance structure are arguably without precedent in a country known for protectionist policies that drain budgets while benefiting narrow elites.

Source: Kuznetsov 2006, adapted from McDermott 2004.

Low-Income Beginners

Active Reformers, Gradual Movers, and Sowers of Seeds

Low-income countries, particularly in Africa, are fighting an uphill battle to join the global economy. Those countries are diverse, however, and their economic performance over the past decade has been varied. Some, showing remarkable dynamism, have chalked up achievements in growth and competitiveness; others lag. To draw out common experiences and lessons, a typology of low-income countries may be useful.

A first category includes countries that can be considered "active reformers." Tanzania, Uganda, and Vietnam, for example, have made significant progress in implementing policies that promote growth and reduce poverty, and the fruits of those efforts have begun to appear in some sectors. Among the active reformers, economic reforms have led to better macroeconomic management; the liberalization of markets and trade has stimulated FDI; and an improved business climate has increased opportunities for the private sector. ICT infrastructure has been strengthened and competitive niches discovered. Such sectoral reforms were not carried out simultaneously. As they move forward, active reformers work to build an efficient innovation system, drawing upon the existing infrastructure and tapping into global knowledge and external resources. Vietnam has been particularly successful in seeking out, acquiring, and adapting global knowledge to local needs (see box 3.1). A climate that stimulates innovation and entrepreneurship can generate further pressure to improve the overall economic framework and to carry out further sectoral reforms—in education, for example, to meet demand for higher qualifications and skills. In this regard, Botswana is a role model for active reformers.

A second category of "gradual movers" includes, for instance, Bangladesh (box 8.7), Ghana, and Senegal, which have a long way to go to complete the reforms required to build the various KE pillars. As a consequence, they have been unable to build durable, competitive knowledge-based sectors, and success stories are scattered. Here, what is needed is a virtuous circle of reforms, in, say, education and innovation, accompanied by greater use of ICTs as a development tool. Sustainable entry points that can provide possibilities for increased job creation and income across the economy also need to be identified. The Bangladesh story illustrates a somewhat chaotic development process in which a competitive textile sector has grown up in a context of mediocre governance—evidence of a typical bootstrapping approach driven by an entrepreneurial society operating largely in the informal sector. Ensuring sustainability of the growth process requires further broad reforms.

A third category of countries can be called "sowers of seeds." Most low-income countries in this category have made virtually no progress toward the knowledge economy over the past decade. Most have struggled with reform; in many, economic performance has languished. Given their relative lack of progress, many of these countries are losing ground in the global knowledge economy. Here, what would help most would be improvement in the overall business and governance climate, together with investment in basic education and telecommunications infrastructure. That prescription, simple as it may appear, takes time under the best of circumstances; without pressure from civil society and the business community, little change occurs. What can be done is to sow seeds—that is, to put in place initiatives of various kinds to serve the basic needs of the population (health, nutrition),

Box 8.7 Bangladesh: Reforms and Industrial Development

Bangladesh has made impressive economic and social gains since the 1990s, despite widely held perceptions of poor governance. Although per capita growth has doubled during the period and long strides have been taken toward some of the Millennium Development Goals (MDGs), Bangladesh scores low on most global governance indicators, and investors rank corruption among their most serious concerns.

How has Bangladesh managed to do so well despite such poor governance? The answer lies in the country's successes and failures. Gains in public accountability have been slow but steady, with successive free elections, an increasingly assertive supreme court, a rapidly growing and active civil society, and a relatively free media. The state has encouraged the emergence of a vigorous private sector through sound macroeconomic management and trade liberalization. Financial sector governance is improving, and successive governments have allocated public resources wisely, emphasizing pro-poor expenditures. Governments also have forged strong partnerships with nongovernmental organizations. On the other side, the governance failures include increasingly unhealthy competition between the two major political parties, the high cost of elections (which feeds public corruption), and an overly centralized state that reduces public accountability in the delivery of services.

To grow, Bangladesh needs to improve its investment climate and empower the poor. It also needs to address the problem of declining exports of ready-made garments since the expiration of the Multi-Fibre Arrangement. The country has a comparative advantage in the production of cheaper standard items, thanks to a high-quality, low-cost labor force and a willingness to meet compliance norms and standards set by buyers in the United States and Europe. The ready-made garment sector employs two million workers, 90 percent of them women. The sector's total exports reached $7.6 billion in 2003–04.

The quotas that had facilitated Bangladesh's growth in this sector for several decades were phased out at the end of 2004, and market competition is becoming keener. Because Bangladesh depends heavily on imports for the raw materials needed to manufacture apparel, it needs to overhaul port facilities, improve infrastructure and transport logistics, and streamline its import regime by removing cumbersome procedures. It also needs to raise productivity in the garment industry through investments in training and technology. It should seek to increase FDI in the sector to take advantage of better management, technology, capital, and market access.

Sources: World Bank 2006a; "Finding Your Niche," *The Economist*, February 27, 2003.

improve education, and exploit niche opportunities. As success stories accumulate, trust and pressure for change grow.

A Focus on Africa

Many countries in Africa are putting in place the building blocks of the knowledge economy. They are strengthening their economic and institutional regime and their infrastructure, meeting the basic conditions for an improved investment climate, and seeking initiatives that can set in motion an overall reform process by illustrating reform's benefits and demonstrating that it can succeed. FDI has created potential opportunities in several countries, as have initiatives to stimulate entrepreneurship through business start-ups. Even in difficult environments one finds pockets of vitality, such as the enterprise clusters discussed in this section. If developed and implemented properly, such opportunities can be—indeed, some have already become—a springboard for wider development opportunities, helping to create virtuous cycles. The main challenge for low-income countries is therefore to achieve some progress while these foundations are being built, partly by adopting the right KE policies.

Economic and institutional regime. Several African countries have seen some measure of progress on this pillar of the KAM, as shown in figure 8.9. Many have achieved a significant degree of political stability and macroeconomic reform that has underpinned progress and development in other areas. More often than not, progress has come by implementing sustained reforms on a few fronts that have borne fruit, for example, by improving the investment climate. Uganda, for

Figure 8.9 *Benchmarking the Economic and Institutional Regime (EIR) in Selected African Countries and Comparators*

Improvement or regression on EIR over period from 1995 to most recent period (2004–05 in most cases)

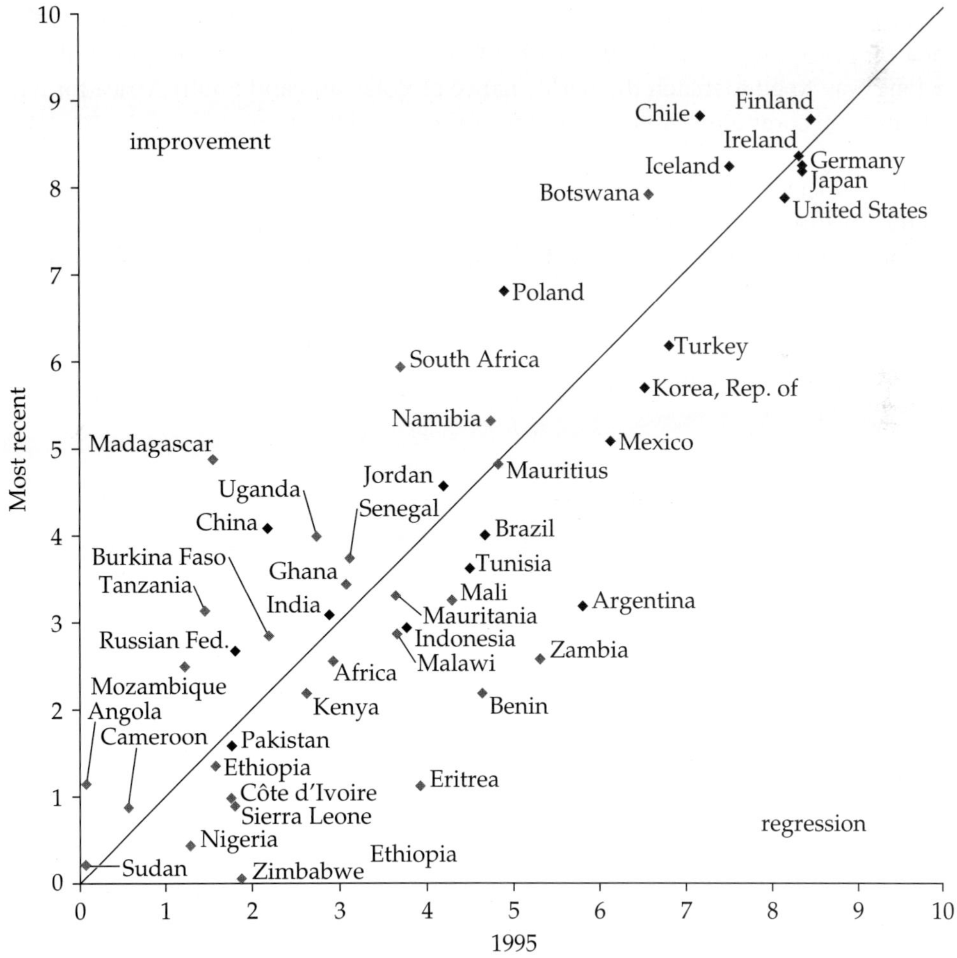

Note: Countries above the 45-degree line have improved their position on the EIR in the most recent period for which data are available relative to their position in 1995 (or the closest available date in the mid-1990s). Countries below the line have weakened their position. The KAM represents default countries in black and countries chosen by the user in blue.

Source: KAM version 2006.

example, recorded robust economic growth averaging 6.8 percent in the period 1990–2003 by carrying out macroeconomic reforms that imposed fiscal discipline, restructured public expenditures, and liberalized the economy. The reform agenda achieved significant improvements in the business and investment climate, notably lower tariff and nontariff barriers and improved rule of law (World Bank 2005g). In Tanzania, the drive to achieve macroeconomic stability through a policy of sound fiscal and monetary management was accompanied by structural reforms, including the privatization of state-owned enterprises, the liberalization of agriculture, improvements to the business environment, and stronger management of public expenditure. As a result, the last five years have seen Tanzania's annual growth rate averaging more than 5 percent, reaching a peak of 6.7 percent in 2004 (Utz 2006).

EDUCATION AND HUMAN RESOURCES. Adult literacy rates in selected African countries are shown in figure 8.10. In tertiary education, many countries' performance is low but beginning to improve (figure 8.11).

INFORMATION AND COMMUNICATION TECHNOLOGIES. Even as African countries have improved their information infrastructure over the last decade, most remain at a very early stage of ICT application and use. Many have made progress but have a long way to go to match the performance of Botswana and South Africa for telephones and computers, for example (figures 8.12 and 8.13).

LOCAL DYNAMICS. In order to grow, countries need to find and build on entry points of innovation. Clusters have emerged in many parts of Africa, often as a spontaneous agglomeration of enterprises and other related factors. Clusters in

Figure 8.10 *Adult Literacy Rate in Selected African Countries, 1990–2004*

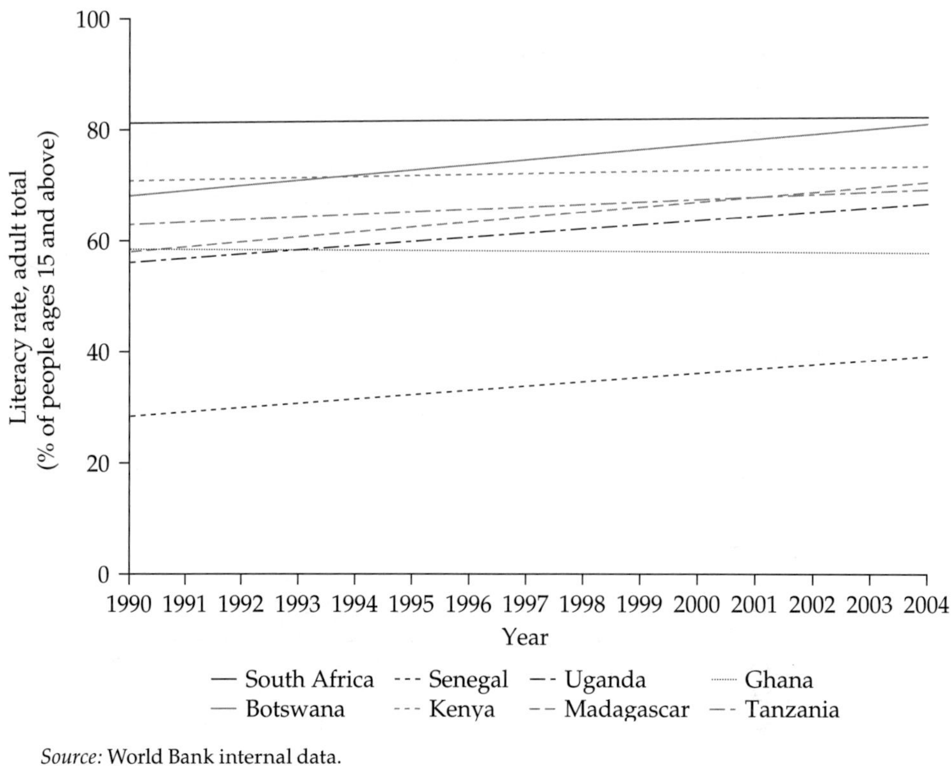

Source: World Bank internal data.

Figure 8.11 *Gross Postsecondary Enrollment Ratio in Selected African Countries, 1991–2004*

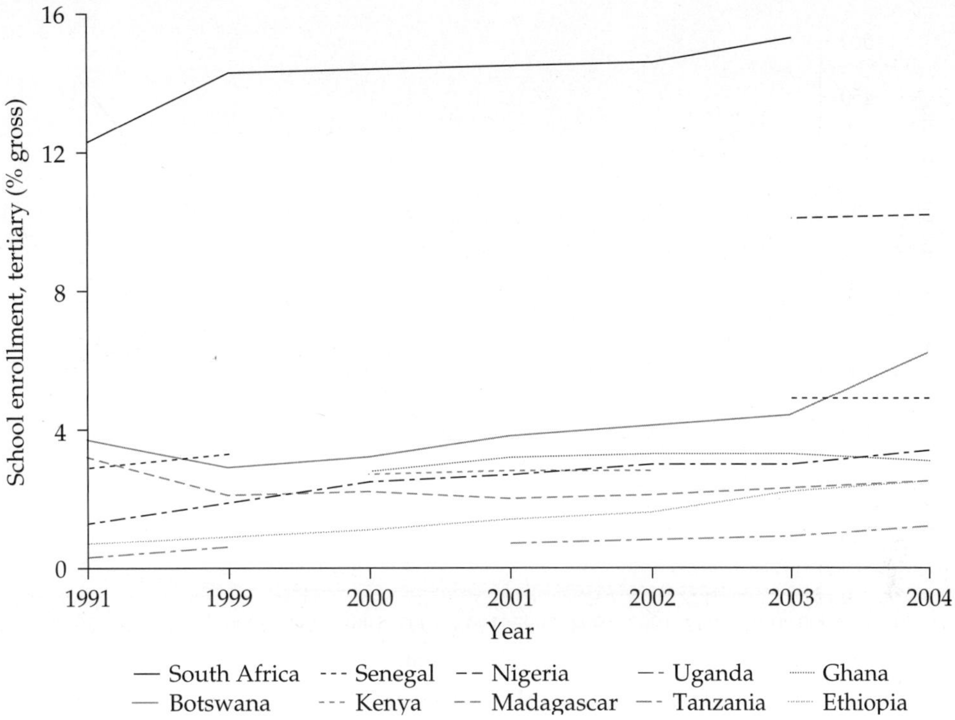

Source: World Bank internal data.
Note: Discontinuous lines are due to unavailability of data.

Ghana, Kenya, Nigeria, Tanzania, and Uganda, for example, centered on natural resources, manufacturing, and high technology (auto parts, computers), have enabled enterprises to overcome binding constraints in capital, skills, technology, and markets to grow and compete (table 8.3). They do so through more effective knowledge and technology diffusion, collective efficiencies, product specialization based on local comparative advantage, and the establishment of production value chains. In some cases the government took limited action to facilitate the growth of a given cluster, but in very few cases did it have an overall cluster policy.

Universities as a source of change. In Francophone Africa, in particular, universities have begun to reexamine their contribution to their countries' development, stimulated by several factors: continuing widespread poverty (and academics' rising concern about it); growing accountability in the allocation of scarce national public resources; competition from higher education institutions abroad and from private ones at home; and the emerging awareness of knowledge as source of economic growth and welfare. Universities are contributing to social and economic development through socially responsive curricula and research agendas, research activities tailored to the needs of local communities, and civic action.[10] By staying

10. These trends are illustrated by examples gathered at a seminar organized by the World Bank Institute, the Agence Universitaire de la Francophonie, and the Francophone Africa University Rectors Conference (CRUFAOCI) in December 2005 in Dakar. For more information, consult the WBI's Knowledge for Development Web site (www.worldbank.org/wbi/knowledgefordevelopment).

Figure 8.12 *Telephones (Mainlines and Mobile Phones) in Selected African Countries, 1990–2004*

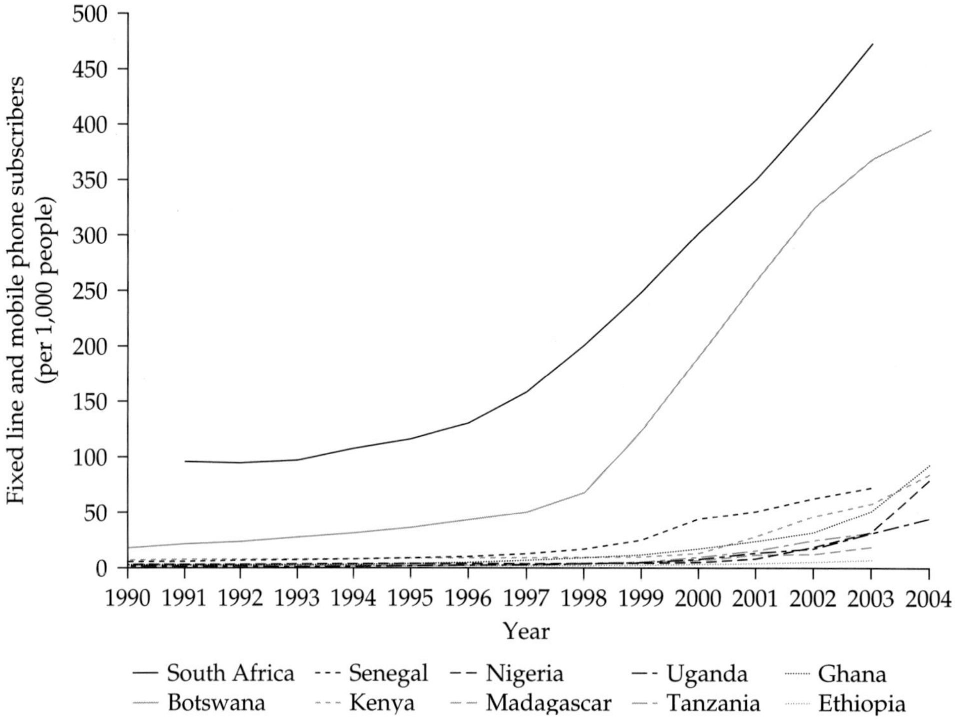

Source: World Bank internal data.

Figure 8.13 *Personal Computers in Selected African Countries, 1990–2002*

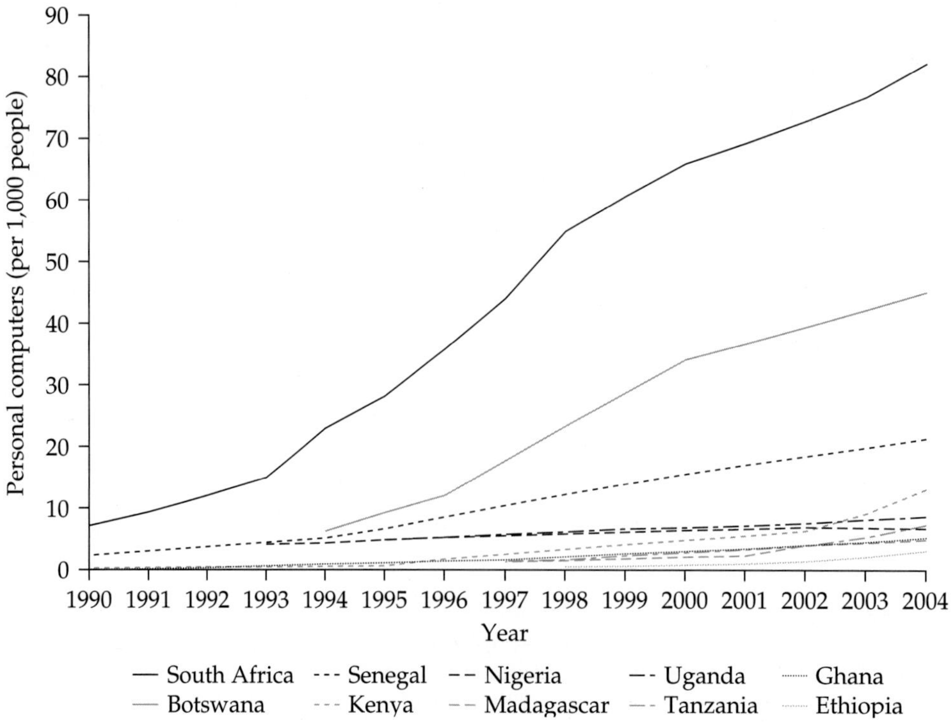

Source: World Bank internal data.

Table 8.3 Eleven Enterprise Clusters in Africa

Country	*Cluster*	*No. of firms*	*Firm size (average no. of employees)*	*Markets*	*Major challenges*
Ghana	The Suame Manufacturing Cluster	> 9,000	5–10	Domestic and limited export (West Africa)	Lack of effective dissemination of R&D results to firms; inadequate physical infrastructure
Kenya	The Kamukunji Metalwork Cluster	> 2,000	1–3	Domestic	Low barriers to entry and congestion of microenterprises; weak linkages with knowledge institutions; weak infrastructure support
Kenya	The Lake Naivasha Cut Flower Cluster	24 (large firms)	250–6,000	Domestic and export (mainly Europe)	Resource depletion and environmental pollution
Nigeria	The Nnewi Automotive Components Cluster	85	< 12	Domestic and limited export	Asian competition and poor public goods
Nigeria	The Otigba Computer Village Cluster	> 5,000	8	Domestic and export (mainly West Africa)	Lack of capital, especially long-term financing; weak infrastructure support; vulnerability to foreign exchange and import duties
Tanzania	The Mwenge Handicrafts Cluster	2,200	15–20	Domestic and limited export	Lack of financing; weak firm capacity; weak public institutions and infrastructure
Tanzania	The Keko Furniture Cluster	—	2–130	Domestic and limited export	Weak public institutions and infrastructure; lack of technological support and access to finance
Uganda	The Lake Victoria Fishing Cluster	17 (fishing plants)	35–200+	Domestic and export (mainly Europe)	Falling fish stock and EU quality standards
Mauritius	The Textile and Clothing Cluster	260	170	Domestic and international	Increasing labor costs; enhanced international competition; low productivity
South Africa	The Wine Cluster	> 340 (wine farms)	—	Domestic and international	Lack of effective marketing/branding strategy and expertise; financial constraints for small producers
South Africa	The Western Cape Textile and Clothing Cluster	327	103	Domestic and international	Increasing labor costs; enhanced international competition; lack of innovation both in product and process

Source: Zeng 2007.

Box 8.8 *Research and Innovation in Francophone Africa—Some Examples*

Seaweed culture in Madagascar: Raising women's revenues while ensuring environmental sustainability. To counter overexploitation of *Euchema* (a variety of algae) stocks, which are used by Biomad, a private company, researchers from the Institute of Marine Science at Toliara University have been able to accelerate its growth. In partnership with the national fishing ministry and Biomad, fishermen have been trained to cultivate *Euchema*. They raised their revenues significantly, while preserving local natural biodiversity.

Reducing economic losses in Côte d'Ivoire through fermented manioc flour. Fermented manioc, widely consumed in Sub-Saharan Africa, was transformed into flour by Cocody University researchers as a way of avoiding losses due to the tuber's rapid degradation. Transformation into flour also improves manioc's quality, cleanliness, and edibility. The innovation was patented in 2000 and has proven its social and economic sustainability and technical feasibility. Financing is needed to advance the process to the stage of industrial production.

Production of essential oil of citronella by rural cooperatives in Togo. The demand for citronella, an insect repellent, is quite large in West Africa. Togo's national productions was low, which provided an opportunity to create a new economic activity for rural populations. A partnership between scientists at Lomé University and international donors led to the commercial distillation of citronella essence by rural cooperatives, now a highly profitable activity.

Source: Authors.

in touch with local realities, researchers can develop partnerships to implement, disseminate, and commercialize their innovations in agriculture and other fields (box 8.8). Supporting many of these developments is the Agence Universitaire de la Francophonie (AUF), an association of some 600 higher education institutions worldwide. AUF encourages member universities to exchange experience and, in so doing, favors institutional capacity building.

Many university initiatives remain enclaves within their broader environment. Linkages with public or private partner institutions are not yet strong enough in most cases to change the national innovation climate. Among the obstacles to the scaling up of initiatives are scarcity of resources, poor infrastructure (which limits academics' access to international research and their ability to share their results and experience), and a weak business environment (which stalls the development of innovations initiated by university teams, as in the case of manioc flour, discussed in box 8.8).

The countries of Sub-Saharan Africa face major challenges in attracting FDI and building strong companies, both of which are critical for sustainable economic and job growth. The International Finance Corporation's (IFC's) Global Business School Network (GBSN) addresses this challenge by focusing on one of the root causes: a lack of adequate management education and training at the local level. GBSN works through local business schools in Kenya, Nigeria, and Ghana, helping to strengthen their programs, sharing best practices, and connecting schools with experts around the world to help them develop a strong pool of management talent for local, regional, and multinational firms and organizations.

Regional and global partnerships. Policy reforms and local initiatives are more likely to succeed and endure if they are integrated with regional and global networks and programs. The work of AUF with the continent's francophone universities and IFC's work with its business schools have already been noted. Two more examples of fruitful regional partnerships follow.

The African Union's New Partnership for Africa's Development (NEPAD) is implementing a noteworthy system for peer review of governance. The voluntary African Peer Review Mechanism (APRM) is making credible strides to improve governance on the continent. A panel of eminent persons appointed by the APRM secretariat oversees the program's implementation throughout Africa. Algeria, Ghana, Kenya, Rwanda, and South Africa have been or are being reviewed

Conceived and designed by African scientists and professionals, the African Institute of Science and Technology (AIST) is a global effort to foster Africa's economic growth and diversification, industrial development, and employment creation by promoting excellence in science and engineering. Eventually AIST will comprise several S&T higher education campuses and smaller affiliated centers of excellence in Africa.

Creating a climate for change and innovation. In the foregoing examples of enterprise clusters, university action, and regional partnerships, the role of the government should be to create a climate favorable to the flowering of initiative and innovation, to publicize success stories and news of new opportunities, to accommodate well-founded and peaceful demands for changes, and to strengthen institutional capability in accordance with a set of priorities designed to remove bottlenecks to growth.

Conclusion: Moving Ahead to a Knowledge Economy

Our concluding messages, drawn from the policy orientations formulated throughout the report, have been grouped under the following action points:

- Putting the knowledge economy (KE) at the heart of development strategies
- Implementing voluntarist and pragmatic KE policies to country specificities
- Acting ambitiously on the four KE pillars
- Making institutions more responsive to KE needs and opportunities.

Putting KE at the Heart of Development Strategies

Countries that are striding toward a knowledge economy—with a solid education base, dynamic information infrastructure, effective innovation system, and solid economic and institutional regime—are doing better than others in the intensely competitive world economy. This is true for countries at all levels of development, particularly for middle-income countries that are catching up with the advanced economies.

The opportunities for countries in the developing world to become successful knowledge-based economies are plentiful, and leapfrogging is a real possibility. If the right policies are adopted, the chances of success are high. Innovative technologies are accessible worldwide, and ambitious innovation strategies that can be used to exploit those technologies at various levels of development can pay off handsomely. The digital divide is narrowing, and returns on economy-wide investment in information and communication technologies (ICTs) can be very high. Education—more and better—is crucial for aspiring knowledge economies. The way to get it is through greater collaboration of the private and public sectors, more effective use of ICTs, continued internationalization of education networks, and a stronger commitment to lifelong learning. An effective and robust economic and institutional regime, conducive to entrepreneurship and attractive to external investors, is fundamental. It is possible to improve segments of this regime in a few years, and even in a few months in certain sectors.

Policy makers must think in KE terms. The ascent of China and India has considerable implications for the international division of labor, as these two increasingly competitive powers neutralize or appropriate comparative advantages once held by other states. KE-oriented strategies and reforms that steer states into suitable niches can create jobs for youths entering labor markets and help society cope better with complex problems, such as clogged cities and environmental damage, which require knowledge and innovation to solve.

Adapting KE Policies to Country Specificities

Achieving KE status is a matter of will and determination, as shown by the success stories presented in this volume. Sometimes serious difficulties, such as an economic crisis, can be a good opportunity for adopting new, KE-inspired strategies. But even without a prompting crisis, *bold visions can inspire enthusiasm across the nation.* To harness that enthusiasm, the vision must be set down in the form of detailed plans for each of the four KE pillars. They should be designed and implemented in such a way as to produce quick and visible results that will build the self-confidence of concerned communities.

KE policy agendas cannot be disconnected from other key policy agendas, such as macroeconomic management and provision of infrastructure services, or from public sector reforms. In fact, the efficiency and sustainability of KE policies will depend in part on other reforms. If those reforms are spotty, the effect of KE policies and investments may be attenuated. Despite these linkages, *KE policies should be implemented somewhat independently of other reform policies,* because KE-related goals, if expressed in a bold and galvanizing way, can act as a powerful lever for other reforms, especially if these are portrayed as prerequisites for the ultimate KE goals.

KE efforts and policies must be adapted to country specificities. Indicators and benchmarking can help identify a country's strengths and weaknesses on the four KE pillars. Governments of countries at different levels of development can then choose good practices related to all four pillars. Policy action must be adapted to the development trajectory, political economy conditions, and sociocultural features of each country.

Although the basic policy principles of the knowledge economy hold at all levels of development, *priorities for action will vary with a country's development situation.* For example, low-income countries must take into account both their needs and capabilities and may have to give priority to the development of basic education and infrastructure for the diffusion of technology. The priority of middle-income countries may be to build competitive and innovative clusters.

As important as bold visions and galvanizing goals may be, *governments must act pragmatically,* taking into account resistance to change and institutional inertia. For that reason, *KE efforts often must begin at the pilot scale.* A critical mass of successful projects will build confidence and remove obstacles to change, clearing the way for major reforms. The process may take several years. Governments should find the right balance between identifying and cultivating dynamic spots in the economy, which may remain enclaves, and society at large, which may need broader actions, such as KE-oriented education reforms, measures to support innovative small and medium enterprises, and an efficient and broadly accessible ICT infrastructure.

Leadership structures, including intermediary bodies linking top KE policy makers with the political, business, and science communities, are essential for successful reform. These bodies need to be involved in formulating vision statements, discussing policies, and evaluating achievements. Communication campaigns often pay off well in marshaling and maintaining popular support for KE initiatives.

Acting Ambitiously on the Four KE Pillars

KE-based growth results from productive interaction between a favorable entrepreneurial climate and wise use of knowledge assets. Among the first steps needed to create an entrepreneurial climate are removal of excessive regulation and all forms of bureaucratic impediments, the establishment of an open economy friendly to trade and foreign direct investment, and measures to ensure that credit is available to small and growing enterprises and communities. Where wholesale reforms to improve the business environment are impossible (for lack of will, consensus, or capacity), governments may first undertake locally focused reforms (such as special export zones), or sector-focused reforms. But experience shows that across-the-board improvements in the business environment *are* possible within a few years.

Innovation is the diffusion of technologies and practices that are new to a given society. Innovation begins with support for dynamic individuals, firms, and communities. That support should be technical, financial, and regulatory—and it should be delivered locally. Each country's research and technology infrastructure must be adapted to the society's needs and capabilities, but it should be managed and funded based on principles of openness and balance between assured (institutional) funding and contingent (contract-based) funding. Cluster growth depends much more on general incentives and a supportive climate characterized by ample supplies of expertise and knowledge than on the provision of specific subsidies and aids.

ICTs are the basic infrastructure of knowledge-based economies. The extraordinary explosion of mobile telephony is facilitating the expansion, access, and use of ICTs in developing countries. But achieving a threshold of fixed-line telephony as a condition for the growth of reliable Internet nodes should not be neglected. The development and use of e-applications in government, business, and education sectors should proceed in tandem with efforts to raise the IT literacy of the population. Comprehensive national programs aimed at attuning citizens to the IT revolution have proven effective even in poor countries.

Knowledge economies usually require profound changes in education systems. Upgrading education in low- and middle-income countries starts with a rapid reduction of illiteracy, and the cultivation of employability skills and self-learning capabilities. Secondary and vocational students must acquire basic competencies, such as reading, writing, reasoning, and basic mathematics. Universities must adjust their programs to world standards; open themselves to the needs of employers, entrepreneurs, and their communities; and develop public-private partnerships. Systems in place in developed countries for efficient lifelong learning—such as short practical courses for obtaining qualifications to meet labor market needs—should inspire developing countries.

Many developing countries can draw on large communities of expatriates, which often include significant numbers of educated people, some of whom have enjoyed very successful careers in developed countries. Expatriates can be a source of expertise, markets, and even venture capital for local entrepreneurs. Sometimes they can be induced to return to their country of origin to help develop new industries. And they can help in the structuring of efficient KE policies and programs.

Making the State and Institutions Responsive to KE Needs and Opportunities

KE strategies are inspired by a vision of the state that differs from models that are solely free-market oriented or solely interventionist. The government should play a strong facilitative role, guiding and encouraging bold development strategies. It should be a policy integrator, bringing synergy to the actions of traditionally separate departments. And it should be a source of stimulation and support for the private sector.

Because knowledge economies turn on productive interactions among people and groups, they depend on effective institutions. The bodies that make and lead KE policies should enlist the participation of all key actors, be positioned at the highest level of government, and act autonomously to set priorities and allocate resources. KE programs should be based on the principle of matching funds, to ensure the full commitment of diverse partners to their funding and implementation. A substantial share of the budget for educational institutions and research institutes should come from contracts with the business community.

Local initiatives are essential to building knowledge economies, because pioneering change often proceeds from such initiatives. Governments should accompany them with decentralization measures bolstered by appropriate reallocation of resources and the establishment of appropriate mechanisms of accountability.

Organizations—research institutes, educational institutions, private enterprises—that thrive in a KE context have both global connections and local ties. Global connections open two-way channels through which flow world knowledge and expertise; local ties ensure that the knowledge economy benefits, and is linked to, the local economy and society. This "glocalization" principle applies to least-developed countries, as well as to the most advanced.

At the international level, collective action is needed to ensure appropriate frameworks for KE trends and country initiatives. The frameworks should facilitate the sharing of science and technology for energy, agriculture, and health; fair trade agreements; adjustments of rules on intellectual property rights; and measures to tackle migration, brain drain, and other human-resource issues. These questions, only briefly evoked in this report, should be examined in depth.

Epilogue: Better Knowledge and Understanding of Society for Promoting Knowledge Economies

As a final word, it is worth emphasizing that better knowledge of the functioning of human societies is a key to the development of knowledge-based economies. It is clear that there is a need for:

- Better self-knowledge among peoples and communities in order to discover how to build trust-based and inclusive societies
- Better understanding of how pressures and incentives affect the evolution of institutional behavior and vested interests
- Better indicators to measure knowledge and innovation achievements, while acknowledging that quantitative measures will never capture completely the complexity of human societies.

Appendix

World Bank Publications on the Knowledge Economy

For more information on the Knowledge for Development Program at the World Bank, please visit www.worldbank.org/wbi/knowledgefordevelopment.

To obtain these and other World Bank publications, visit www.worldbank.org. All volumes are published by the World Bank unless otherwise noted. Many of the works listed here are available free of charge in PDF format.

The Concept of the Knowledge Economy

- World Bank. 1999. *World Development Report 1998/99: Knowledge for Development.* Oxford University Press.

At the World Bank, work related to the knowledge economy began with the publication of *World Development Report 1998/99: Knowledge for Development.* This report showed that despite the cost of creating new knowledge—most of which is borne by industrial countries—developing countries can compete by borrowing knowledge from abroad, in addition to creating their own where advantageous. The report used Ghana and the Republic of Korea to make its point: 40 years ago, both countries had virtually the same income per capita; but by the early 1990s, Korea's income per capita was six times that of Ghana. This can be largely attributed to Korea's greater success in acquiring and using knowledge. The report also addressed the twin issues of knowledge gaps and information problems and emphasized that governments must recognize and respond to both types of problems simultaneously to unleash the power of knowledge

Knowledge and Growth

- Derek H.C. Chen and Carl J. Dahlman. 2004. "Knowledge and Development: A Cross-Section Approach." World Bank Policy Research Working Paper 3366.

Chen and Dahlman assessed the effects of different aspects of knowledge on economic growth. Using an array of indicators, each of which represented an aspect of knowledge, as independent variables in cross-sectional regressions spanning 92 countries from 1960 to 2000, the authors showed that knowledge is a significant determinant of long-term economic growth. In particular, the stock of human capital and levels of domestic innovation, technological adaptation, and information and communications infrastructure all exert statistically significant positive effects on long-term economic growth. An increase of 20 percent in the average years of schooling of a population increases average annual economic growth by 0.15 percent. A 20 percent increase in the annual number of U.S. patents granted is associated with an increase of 3.8 percent in annual economic growth. And when the information and communications infrastructure, measured by the number of phones per 1,000 persons, is increased by 20 percent, annual economic growth tends to increase by 0.11 percent.

Innovation

- Aubert, Jean-Eric. 2005. "Promoting Innovation in Developing Countries—A Conceptual Framework." World Bank Policy Research Working Paper 3554.

Aubert provides a conceptual framework for the promotion and diffusion of technological innovation in developing countries. The innovation climate in developing countries is plagued by poor business and governance conditions, low educational levels, and mediocre infrastructure. Promotion of innovation should be understood as the diffusion of technologies and related practices that are new in a given context (not in absolute terms). It is necessary first to provide a package of support—technical, financial, commercial, legal—delivered by flexible, autonomous agencies that adapt their support and operations to the different types of concerned enterprises. Facilitating and responding to local needs is also essential. At the global level, issues such as the role of foreign direct investment (FDI), technology patenting and licensing, and the North-South research asymmetry need to be considered and redressed through appropriate incentives and regulations.

- United Nations Millennium Project. 2005. Investing in Development—A Practical Plan to Achieve the Millennium Development Goals. Available from http://www.unmillenniumproject.org/reports/index_overview.htm

To promote innovation, development policies must focus on key sources of economic growth, particularly the use of scientific and technological knowledge. The United Nations Millennium Project outlines core areas for policy action: (a) focusing on platform or generic technologies, (b) defining infrastructure services as foundations for technology, (c) placing universities at the center of local development, (d) improving science education, (e) spurring entrepreneurial activities, (f) improving the policy environment, and (g) developing areas of underfunded research.

- Robert Watson, Michael Crawford, and Sara Farley. 2003. "Strategic Approaches to Science and Technology in Development." World Bank Policy Research Working Paper 3026.

Watson, Crawford, and Farley (2003) examine the ways in which science and technology (S&T) support poverty alleviation and economic development and how these themes have been treated in the World Bank's work. The paper presents policy options for enhancing S&T systems in developing countries, reviews the previous experience of the World Bank and other donors in supporting S&T, and suggests changes that the World Bank and its partners can make to improve current work. Its main message is that while S&T have always been important for development, the unprecedented pace of advancement of scientific knowledge is creating challenges to development, for which most developing countries are unprepared. The World Bank could encourage the needed capacity improvements in client countries by devoting greater attention to S&T in education, health, rural development, private sector development, and the environment.

- World Bank. 2004. "Innovation Systems: World Bank Support of Science and Technology Development." World Bank Working Paper 32.

Innovation systems and S&T projects supported by the Bank have taken many forms in the past several years. This paper reviews the lessons learned in Bank-supported S&T projects from 1989 to 2003. Because Bank experience is most substantial in innovation systems and related policy frameworks, the paper focuses on industrial technology development and building national innovation systems.

- Daniel Lederman and Laura Saenz. 2005. "Innovation and Development Around the World: 1960–2000." World Bank Policy Research Working Paper 3774.

Lederman and Saenz examine innovation and development around the world from 1960 to 2000, presenting a database of innovative activity. Their chief indicator of innovation output is patents. The variables of innovation input are investment in R&D and technical personnel (engineers, scientists) working in R&D activities. The authors examine trends and patterns of innovation output and input by studying the movement over time of the relevant series and comparing the performance of developing and high-income countries. They also provide cross-regional comparisons and a detailed examination of trends in selected countries. The econometric results suggest that innovation might indeed have strong positive effects on long-run development, even stronger than the direct effects of institutions.

- Mariano Bosch, Daniel Lederman, and William F. Maloney. 2005. "Patenting and Research and Development: A Global View." World Bank Policy Research Working Paper 3739.

Bosch, Lederman, and Maloney apply econometric techniques to global data on patents and innovation inputs to explore the dynamic relationship between R&D and U.S. patents granted. Their work affirms the recurrent micro-level finding of a strong relationship between the two at the national level. This conflicts with the frequent micro-level findings of strong diminishing returns to knowledge generation and suggests the importance of spillover effects measurable only at the aggregate level. Developing countries, however, do show diminishing returns, and implicit rates of return to innovative effort in these countries appear to be a fraction of those found in the industrialized world. Across the entire sample, several variables—education, intellectual property rights, and (in some specifications) the quality of research institutions and their interaction with the private sector—affect the elasticity of transformation of R&D into patents.

- Vandana Chandra, ed. 2006. *Technology, Adaptation, and Exports: How Some Developing Countries Got it Right.*

Chandra and her collaborators study ten cases of developing countries (including Kenya and Uganda) that have successfully adopted new technologies to catch up with developed countries in a particular industry. These examples demonstrate that latecomers can leapfrog across several stages of technological development. The study covers a range of industries in which learning how to use and upgrade new technologies was the driver of development. The book highlights the institutional structures and policies that allowed countries to acquire, adapt, and disseminate technologies to hone their export competitiveness. Although Chandra's volume focuses on technological adaptation, the cases make clear that a stable macroeconomic, political, and social environment supported by a legal system that enforces contracts and protects property rights is also essential for catch-up.

Information and Communication Technologies

- World Bank. 2003. "ICT and MDGs: A World Bank Group Perspective." Working Paper 27877.

During the past decade, the international community has focused on strategies to help the people of the world's poorest countries share in the benefits of globalization and escape the traps of poverty. In 2000, the G-8 initiated the Digital Opportunity Task Force (DOT Force) to harness new technologies that can narrow social and economic inequalities. This study reports that the use of information and communication technologies (ICTs) to achieve the Millennium Development Goals (MDGs) is still in its early stages, but the seeds of success are being sown across the globe. The document illustrates the opportunities ICT offers policy makers and practitioners in their efforts to achieve the MDGs and highlights selected projects funded by the World Bank Group that are using ICTs to accelerate development.

- Kerry S. McNamara. 2003. *Information and Communication Technologies, Poverty, and Development: Learning from Experience.*

In this background paper for the infoDev Annual Symposium in Geneva in December 2003, Kerry McNamara offers a framework for ideas about ICT, poverty, and development that could guide further analysis and future projects. Its focus is on ICT as a tool to change the performance of institutions and markets, the livelihood of the poor, and the capacity of individuals and governments. These changes, not ICT itself, lead to poverty reduction and sustainable development.

- World Bank. 2006. *Information and Communications for Development 2006: Global Trends and Policies.*

This annual flagship publication takes stock of worldwide progress in widening access to ICTs. The report, which includes ICT indicators for almost 150 countries, builds on experience drawn from the Bank's own significant involvement in the sector. Despite greater access, however, poor countries still lag behind in ICT applications in governments, schools and business. The report urges the governments of developing countries to work across ministries and to enter into partnerships with the private sector to extend the reach of ICTs.

- World Bank. 2005. *E-strategies: Monitoring and Evaluation Toolkit.*

The Bank has also focused on e-strategies for development. The essential point of this toolkit from the Global Information and Communication Technologies department (a joint World Bank–International Monetary Fund unit) is that monitoring and evaluation (M&E) are integral to the design and implementation of effective e-strategies. Developing the M&E components of e-strategies ensures that the latter will be explicit about and realistic in their aims, and that their implementation will be regularly assessed and realigned to conserve scarce resources. In many respects, the credibility and efficiency of e-strategies depend on whether they have a strong M&E spine. Based on a review of some 50 e-strategies, the authors propose a framework for use in integrating M&E into e-strategies.

- World Bank. 2005. *E-Development: From Excitement to Effectiveness.*

This complementary volume, edited by Robert Schware for the World Summit on the Information Society in Tunis in November 2005, examines the requirements and realities of using ICTs to advance development goals. Rather than trying to present a comprehensive overview of e-development, the volume highlights key issues that have immediate relevance to policy makers in developing nations who make decisions on investments and development goals. It highlights two issues: e-government and e-education, because ICT applications in these areas can lead to significant development, and they can also be successfully deployed through public-private partnerships, leveraging limited government funding to achieve greater impact.

Education

- World Bank. 2005. *Expanding Opportunities and Building Competencies for Young People: A New Agenda for Secondary Education.*

In a knowledge-based economy, it is not enough to invest in primary education. Countries wishing to produce a cadre of well-educated people who can access, create, and use knowledge must provide greater access to secondary education and tertiary education to have. The World Bank has recognized the role of secondary education in fostering knowledge-based development. This study emphasizes that knowledge and skills hold the key to a gainful and productive future. Countries must move from elitism to inclusiveness, providing all citizens the same chance at secondary learning, with the university and the job market being distinct possibilities for the future. The report sets forth evidence-based, "tried and true" policy options for decision makers, intended to help developing countries and transitioning economies adapt their secondary education systems to the demands created by the expansion of primary education, the social and economic challenges presented by globalization, and the fast pace of the knowledge-based economy.

- World Bank. 2000. *Higher Education in Developing Countries: Peril and Promise.*

In 2000, the World Bank and UNESCO convened a task force that brought together experts from 13 countries to explore the future of higher education in developing countries. Based on research, discussion, and hearings conducted over a two-year period, the task force concluded that, without better higher education, developing countries will find it increasingly difficult to benefit from the global knowledge economy.

- World Bank. 2002. "Constructing Knowledge Societies: New Challenges for Tertiary Education." Education Group, Human Development Network.

Even though the World Bank has supported tertiary education reform efforts for many years, some critics have alleged that the Bank has not been responsive enough to the growing demand by clients for tertiary education interventions and that lending has not matched the growing importance of tertiary education for economic and social development, especially in poor countries. This report describes the role of tertiary education in building up a country's capacity for participation in the knowledge economy and investigates policy options that may enhance economic growth and reduce poverty.

- Jamil Salmi and Arthur M. Hauptman. 2006. "Innovations in Tertiary Education Financing: A Comparative Evaluation of Allocation Mechanisms." Education Sector Working Paper 4.
- William Saint. 2006. "Innovation Funds for Higher Education: A Users' Guide for World Bank Funded Projects." Education Sector Working Paper 1.

Salmi and Hauptman examine recent experiences with both traditional and more innovative allocation mechanisms for tertiary education worldwide. Their work surveys the landscape of innovation and highlights lessons that can help policy makers in developing and transitioning countries formulate their public policies for tertiary education. William Saint has developed a users' guide for innovation funds in tertiary education that is intended for Bank professionals who seek to cultivate demand-driven processes for improvement of educational quality.

- World Bank. 2003. *Lifelong Learning in the Global Knowledge Economy: Challenges for Developing Countries.*

The emergence of the global knowledge economy has put a premium on lifelong learning. The World Bank's report on the subject is an attempt to lay out an analytical framework for understanding the challenges of developing a lifelong learning system. The World Bank's involvement in lifelong learning is still at the early stages, but some projects—in Romania and Chile, for example—address the need for continuing education and lifelong learning.

- Carl Dahlman, Douglas Zeng, and Shuilin Wang. 2007. *Building a System of Life-Long Learning to Enhance China's Competitiveness.*

To extend its analytical work on lifelong learning and begin a policy dialogue that touches more closely on the issues, the Bank prepared a report on lifelong learning in China. The report outlines the key elements of an integrated lifelong learning system, with economic aspects the primary focus. It presents a framework for understanding the demands on the education and training system, and describes several policy thrusts for building an effective lifelong learning system. The role of government would evolve from that of provider to one of architect, facilitator, and rule-keeper. In its new role, the government would ensure quality, relevance, efficiency and equity through a sound accreditation, assessment, and vocational qualification system based on strong linkages with the labor market. While establishing financial aid programs, it would also foster the growth of an education finance market and tap into private resources to meet the increasing demand for education and training. And it would create an environment within which the potential of distance education could be exploited.

Regional KE Reports

Latin America and the Caribbean

As the gap between knowledge-based and labor-based economies grows wider, many countries in the Latin America and Caribbean (LAC) region have found themselves on the losing end of the divide. Why are so many countries in the region losing ground in the knowledge economy?

- World Bank. 2002. From Natural Resources to the Knowledge Economy: Trade and Job Quality.

To answer some of these questions, in its first flagship report, the LAC region at the World Bank addresses three concerns about the structure of trade in LAC economies. The first is whether natural wealth and exports of natural-resource-intensive commodities hamper economic development. The second concern is that natural resources create a concentrated export structure that exacerbates economic volatility and thus reduces growth. The third concern is that international trade might eliminate jobs. The wide-ranging report cites the experience of Australia, Canada, Finland, Sweden, and the United States, as well as some Latin American countries, to show how successful economies have been built on the basis of primary commodity exports.

- World Bank. 2003. *Closing the Gap in Education and Technology.*

The region's second flagship report looks at how investment in education, opening up to new technologies through foreign trade and investment, and encouraging private sector R&D are the keys to harnessing technology to speed growth in the region. Latin America's income growth has lagged because of a "productivity gap" caused by the region's failure to keep pace with new technologies in its production processes and skills. To close the productivity gap, the report calls for a range of policy approaches and strategies, depending on a country's level of development. This study identifies three progressive stages in a country's technological evolution—adoption, adaptation, and creation—and observes that policies should be designed to address the particular challenges that accompany each stage.

- World Bank. 2007. "Innovation Policies in Latin America II." Office of the Chief Economist, Latin America and the Caribbean Region.

A recently completed study finds that differences in total factor productivity explain roughly one-half of the difference in per capita GDP across the region. Many believe that Latin America faces an innovation problem, often manifested as a low R&D investment rate. This study has two goals: to put the diagnostic of the innovation shortfalls on a stronger analytical base, and to contribute to the policy dialogue by offering conceptual and empirical support for different types of interventions. Given the region's concern with government capacity, this study's aim is to identify the relevant market failure and the interventions and incentives needed to remedy it.

East Asia and the Pacific

- Shahid Yusuf and Simon J. Evenett. 2002. *Can East Asia Compete?* World Bank and Oxford University Press.
- Shahid Yusuf. 2003. *Innovative East Asia: The Future of Growth.* Oxford University Press.
- Shahid Yusuf, M. Anjum Altaf, and Kaoru Nabeshima, eds. 2004. *Global Change and East Asian Policy Initiatives.* World Bank and Oxford University Press.

- ———. *Global Production Networking and Technological Change in East Asia.* World Bank and Oxford University Press.
- Shahid Yusuf and Kaoru Nabeshima. 2006. Postindustrial East Asian Cities: Innovation for Growth.

Since 1999, the Bank has conducted several studies in East Asia to identify the most promising path to development in light of emerging global and regional changes. The first volume, *Can East Asia Compete?*, provided a compact overview of the relevant strategic issues and future policy directions. A year later, *Innovative East Asia* analyzed the main issues and consequent policy choices, drawing comprehensively from recent empirical research and the findings of firm surveys conducted for the study. Its main message was that sustained economic growth in East Asia will depend on the ability to retain the strengths of the past—stability, openness, investment, and human capital development—to overcome the sources of current weaknesses—the financial, corporate, judicial, and social sectors—and to implement the changes required by the evolving economic environment, particularly in technology. In 2004, the Bank published *Global Production Networking and Technological Change in East Asia*, which marshaled detailed information, analysis, and case studies to show that the economies of East Asia must adapt to the changing character of global production networks and nurture their technological capabilities in order to sustain growth. *Global Change and East Asian Policy Initiatives*, published in 2004, argued that Chinese firms should pursue strategies that will promote efficiency and innovation to compete in the global market. Future growth in East Asia will increasingly come from the strength of innovative activities instead of factor accumulation. Such innovative activities—especially in producer services and the creative industries—are concentrated in high-tech clusters in globally linked cities. *Postindustrial East Asian Cities*, the most recent work in the series, explores the growth prospects of six cities: Bangkok, Beijing, Seoul, Shanghai, Singapore, and Tokyo. The development of these cities is influenced by ongoing structural changes and initiatives by governments and firms. The authors draw on a wide literature and on interviews with firms conducted especially for the study, which suggests how policies and institutions can yield an urban environment that supports innovative activities.

Sub-Saharan Africa

- World Bank. 2000. *Can Africa Claim the 21st Century?*

The World Bank's Africa Region has done some important studies on growth. *Can Africa Claim the 21st Century* points out that many African countries have made important economic reforms—where these reforms have been sustained, they have raised growth and incomes and reduced poverty. But the response has not been sufficient to overcome years of falling income or to reverse other adverse legacies, including deteriorated capacity, weakened institutions, and inadequate infrastructure. Africa thus faces an immense, multifaceted development challenge. Fortunately, globalization and new technology, especially ITCs, offer great potential for Africa.

- Benno J. Ndulu, L. Chakraborti, and L. Lijane. 2007. *Challenges of African Growth: Opportunities, Constraints and Strategic Directions.*

The authors trace the opportunities, constraints, and strategies for Africa and its partners as they try to accelerate economic growth, reduce poverty, and put Africa on a path to achieve the MDGs. Ndulu draws lessons from 45 years of experience in Africa and around the world and uses them to shape growth strategies. His work addresses three key points: the opportunities for growth available to the African countries; the constraints to exploiting these opportunities; and the choices to be made by African governments and their development partners, including the World Bank, in supporting actions taken by African countries. The distinguishing characteristics of the study are its long-term perspective, and its description and analysis of the African growth experience from 1960 (when most African countries became independent) to the present.

- World Bank. 2004. *Indigenous Knowledge: Local Pathways to Global Development.*

This volume, which marked the fifth anniversary of the World Bank Indigenous Knowledge for Development Program, synthesizes lessons learned from the indigenous practices of African communities. Case studies demonstrate how communities and local practitioners use indigenous knowledge and practices to increase crop yields, educate their children, reduce suffering from HIV/AIDS, decrease infant and maternal mortality, heal the impact of conflict, learn from each other, and empower themselves. The cases also suggest that African communities are quite willing, indeed eager, to combine global knowledge and modern technology with their indigenous knowledge and institutions to improve their lives.

- Douglas Zhihua Zeng, ed. 2007. *Knowledge, Technology and Cluster-Based Growth in Africa.*

Although African enterprises face many obstacles, pockets of economic vitality can be found, often in the form of enterprise clusters. These clusters contribute significantly to growth and competitiveness on several levels—national, regional, and local economic growth—usually through more effective diffusion of knowledge and technology. They have enabled enterprises to overcome binding constraints in the areas of capital, skills, technology, and markets by tapping into the global knowledge and technology stocks, leveraging local comparative advantage, fostering value chains, and yielding efficiency gains. In 13 case studies in 7 African countries, this book examines how dynamic enterprise clusters were formed and evolved; how knowledge, human capital, and technology have contributed to their success; and the challenges and constraints they confronted while moving forward. Based on the cases, Zeng proposes some general policy recommendations on how to scale up and sustain these and similar clusters.

Other Regions

- Jean-Eric Aubert and Jean-Louis Reiffers. 2003. *Knowledge Economies in the Middle East and North Africa: Toward New Development Strategies.*

Other regional vice presidencies at the World Bank have also begun innovative work on the knowledge economy. Aubert and Reiffers (2004) analyze the development of knowledge-based economies in the Middle East and North Africa. To date, investments in education, information infrastructure, R&D, and innovation have been insufficient in most countries of the region. Moreover, inadequate economic and institutional frameworks prevent those investments from yielding desired results. The region risks falling farther behind in the world economy unless urgent action is taken to advance structural reform and increase knowledge-related investments.

- World Bank. 2006. "Public Financial Support for Commercial Innovation." ECA Knowledge Economy Study, Part I, Private and Financial Sector Development Department.

The Europe and Central Asia (ECA) region of the World Bank has studied innovation, focusing on public financial support for commercial innovation. World Bank 2006 reviews institutions and conditions in various ECA countries to determine which are ready for public support for commercial innovation and which are missing some critical institutional requirements. The appropriateness of financial instruments used elsewhere to encourage private sector innovation is evaluated. The aim of the recommended financial instruments is to encourage private R&D in companies by providing incentives for collaboration through the co-funding of consortia of firms and research institutes (or universities) to implement innovative projects. Nonfinancial instruments, such as business support services, incubators, and economic support zones, are complementary components. The focus of this study on R&D and commercialization is consistent with the view that commercial innovation and R&D are key factors driving self-sustained, long-term economic growth, and that these factors are generated from within the economic system, responding to economic incentives.

KE Country Studies

- Carl Dahlman and Thomas Andersson, eds. 2000. *Korea and the Knowledge-Based Economy: Making the Transition*. World Bank/OECD.

In *Korea and the Knowledge Economy*, Dahlman and Andersson review the economic transition of the Republic of Korea, focusing on the challenge of the knowledge revolution to the country's development strategy. Despite the country's success, areas of relative inefficiency remain— namely, inadequate conditions for generation of knowledge and information, insufficient competition, and misallocation of investments. The Korean government should redefine its role to unleash the creative power of markets, provide a legal and regulatory framework for more competitive markets, and foster policies conducive to enterprise development, while also addressing the risks of the "digital divide."

- Carl Dahlman and Jean-Eric Aubert. 2001. *China and the Knowledge Economy: Seizing the 21st Century.*

Dahlman and Aubert argue that in order to prosper in the era of globalization, China must make effective use of knowledge in agriculture, industry, and services. China also needs to transition to an environmentally sustainable economy that better manages its relatively limited natural resources. The book presents a long-term strategy for the country that integrates key knowledge-related policy components, such as improving relevant economic incentives and institutions, upgrading the education and training system, building the information infrastructure, and strengthening the innovation and research system. The book recommends that the government withdraw further from hands-on management of the economy and take on the role of architect of appropriate institutions and provider of incentives to establish a new socialist market economy based on knowledge.

- Alfred Watkins. 2003. "From Knowledge to Wealth: Transforming Russian Science and Technology for a Modern Knowledge Economy." World Bank Policy Research Working Paper 2974.

Watkins points out that even though the Russian Federation has world-class basic research capacity and possesses a sophisticated S&T infrastructure, its exports are primarily raw materials—the country lacks an effective system for converting its scientific capacity into wealth. Adjusting the S&T system to the demands of a market economy will require a major program of institutional and sector reform. Watkins describes the status of the Russian S&T sector after 10 years of transition and the evolution of the Russian system of intellectual property rights from Soviet times to the present. He argues that Russia will never develop a successful commercialization program until it clarifies the ownership of the large stock of intellectual property, most of which was funded from federal budget resources. Watkins outlines a comprehensive, 10-point, sector reform program to improve the efficiency of government research and development spending and to link the Russian S&T system with market forces.

- Carl Dahlman and Anuja Utz. 2005. *India and the Knowledge Economy: Leveraging Strengths and Opportunities*. World Bank Institute.

India has the potential to raise productivity by shifting labor from low productivity and subsistence activities in agriculture, informal industry, and informal service to more productive modern sectors, as well as to new knowledge-based activities. To be competitive in the global knowledge economy, India should focus its efforts on reforming further its economic and institutional environment and improving its trade and investment climate. For India to leverage its strengths and opportunities on a global scale, it must undertake significant reforms and investments to improve education and skills, strengthen its innovation system, and bolster its information infrastructure.

- L. Alan Winters and Shahid Yusuf. 2007. *Dancing with the Giants: China, India, and the Global Economy.*

China and India are increasingly important in global markets, and their rapid expansion is having a growing effect on the global commons. Forthcoming World Bank research (Winters and Yusuf 2007) explores whether the present rate of growth can be sustained through 2020, what turns the path of growth might take, how the two countries' expansion might encroach on other countries, whether growth might not be constrained by weaknesses in governance, growing inequality, and environmental stresses. The authors conclude that growth will not be constrained, but they do suggest that the Chinese and Indian authorities face challenges in keeping their investment climates favorable, their inequalities at levels that do not undermine growth, and their air and water quality at acceptable levels. The authors also consider China's and India's interactions with the global trading and financial systems and their impact on the global commons, particularly with regard to climate change.

- Anuja Utz. 2006. "Fostering Innovation, Productivity, and Technological Change: Tanzania in the Knowledge Economy." World Bank Working Paper.

Utz maintains that the prospects for sustained growth that makes the most of Tanzania's recent macroeconomic stability and structural reforms will depend on the economy's capacity for innovation. Utz argues that the quality of education, particularly post-primary education, is crucial for innovation. In the Tanzanian context, innovation refers to products that have not yet been produced successfully in Tanzania and the adoption of technologies and processes new to the country. Thus defined, innovation is the path to economic diversification and moving up the value chain. Although Tanzania has been able to benefit from the ICT revolution, more work is needed to modernize telecommunication policies to generate fair competition and reduce high communication costs. Improvement in these areas will have an important payoff for Tanzania.

- Carl Dahlman and Yevgeny Kuznetsov. 2007. *Mexico's Transition to a Knowledge-Based Competitiveness: Challenges and Opportunities.* World Bank Institute.

Dahlman and Kuznetsov argue that the knowledge revolution offers unprecedented challenges and opportunities for Mexico. To take advantage of new knowledge-based opportunities, however, Mexico must take concerted action now. Mexico should create institutional capabilities through innovation and enterprise upgrades, improve education and skills, and update its ICT infrastructure. Momentum for change must come simultaneously from the bottom up and from the top down. The best way to move forward is to make decision makers aware of the need for consensus on how to overcome national obstacles. Then, Mexico can move ahead with concrete and manageable bottom-up approaches that promote national vision and build on past successes. Concerted action among stakeholders is vital for progress. States and champions in the private sector are important agents of the transition to knowledge-based competitiveness.

Bibliography

Acemoglu, Daron, Simon Johnson, and James A. Robertson. 2000. "The Colonial Origins of Comparative Development: An Empirical Investigation." *American Economic Review* 91 (5): 1369–1401.

Acemoglu, Daron, Simon Johnson, James A. Robinson. 2001. "An African Success Story: Botswana." MIT Working Paper, Massachusetts Institute of Technology, Cambridge, MA. July.

Adams, Arvil V., and Richard K. Johanson. 2004. "Skills Development in Sub-Saharan Africa." World Bank, Washington, DC.

Adams, James D. 1990. "Fundamental Stocks of Knowledge and Productivity Growth." *Journal of Political Economy* 98 (4): 673–703.

ADEA. 2005. "Learning, but in Which Language?" *ADEA Newsletter* 17 (2).

Aggarwal, A., R. Devane, Y. Kuznetsov, and A. Pandey. 2004. "India's Transformation to Knowledge-based Economy: Evolving Role of the Indian Diaspora." Evalueserve, Gurgaon (New Delhi).

Andersson, Thomas, Sylvia Schwaag Serger, Jens Sörvik, Emily Wise Hansson. 2004. *The Cluster Policies Whitebook*. Malmo, Sweden: IKED.

Aubert, Jean-Eric. 2005a. "Knowledge Economies: A Global Perspective." In *Intellectual Capital for Communities—Nations, Regions, and Cities*. New York: Elsevier.

———. 2005b. "Promoting Innovation in Developing Countries—A Conceptual Framework." World Bank Policy Research Working Paper 3554. Washington, DC: World Bank

———. 2006. "Intellectual Capital, Economic Development, and the Cultural Factor." Presentation at Second World Conference on Intellectual Capital for Communities, June 29–30, Paris.

Aubert, Jean-Eric, and Derek Chen. Forthcoming. "The Island Factor as a Growth Booster." *Journal of Intellectual Capital*.

Aubert, Jean-Eric, and Jean-Louis Reiffers. 2003. *Knowledge Economies in the Middle East and North Africa*. WBI Learning Resources Series. Washington, DC: World Bank.

Aubert, Jean-Eric, and Josée Landrieu. 2004. *Vers des civilisations mondialisées—De l'éthologie à la prospective*. Paris: Editions de l'Aube.

Autor, D. H, F. Levy, R. J. Murnane. 2003. "The Skill Content of Recent Technological Change: An Empirical Exploration." *Quarterly Journal of Economics* 118 (4): 1279–1333.

Ayyagari, Meghana, Asli Demirguc-Kunt, amd Vojislav Maksimovic. 2006. "How Important Are Financing Constraints? The Role of Finance in the Business Environment." World Bank Policy Research Working Paper 3820. Washington, DC.

Barro, Robert J. 1991. "Economic Growth in a Cross Section of Countries." NBER Working Paper 3120, National Bureau of Economic Research, Cambridge, MA.

Barton, John H. 2003. "New International Arrangements in Intellectual Property and Competition Law." In Ove Granstrand, ed., *Economics, Law and Intellectual Property: Seeking Strategies for Research and Teaching in a Developing Field* (109–121). Boston, MA: Kluwer Academic Publishers.

Benavente, José-Miguel, Luiz de Mello, and Nanno Mulder. 2005. "Fostering Innovation in Chile." Economics Department Working Papers, Organisation for Economic Co-operation and Development, Paris.

Bhojwani, H. R. 2006. "Report on the Indian Civilian Public R&D System." Background paper prepared for the World Bank Environment for Innovation Study. Washington, DC.

Bosch, Mariano, Daniel Lederman, and William F. Maloney. 2005. "Patenting and Research and Development: A Global View." World Bank Policy Research Working Paper 3739. Washington, DC.

Brinkley and Lee. 2006. "The Knowledge Economy in Europe: A Report Prepared for the 2007 EU Spring Council." The Work Foundation. October. http://www.theworkfoundation.com/Assets/PDFs/KE_Europe.pdf.

Chandra, Vandana. 2006, *Technology, Adaptation, and Exports: How Some Developing Countries Got It Right*. Washington, DC: World Bank.

Chellaraj, Gnanaraj, Keith E. Maskus, and Aaditya Mattoo. 2005. "Skilled Immigrants, Higher Education, and U.S. Innovation." In Maurice Schiff and Caglar Ozden, eds., *International Migration, Remittances, and the Brain Drain* (chapter 8). Washington, DC: World Bank.

Chen, Derek H. C., and Carl J. Dahlman. 2004. "Knowledge and Development: A Cross-Section Approach." World Bank Policy Research Working Paper 3366. Washington, DC.

Chidamber, Shyam. 2003. "An Analysis of Vietnam's ICT and Software Services Sector." *Electronic Journal on Information Systems in Developing Countries* 13 (9): 1–11.

Cincera, Michele, and Bruno Van Pottelsberghe de la Potterie. 2001. "International R&D Spillovers: A Survey." *Brussels Economic Review/Cahiers Economiques de Bruxelles* 169: 3–31. Department of Applied Economics, Free University of Brussels.

CIPR (Commission on International Property Rights). 2002. *Final Report: UK Commission on Intellectual Property Rights*. London.

Cohen, Daniel, and Marcelo Soto. 2001. "Growth and Human Capital: Good Data, Good Results." CEPR Discussion Papers 3025, Centre for Economic Policy Research, London.

Dahlman, Carl, and Anuja Utz. 2005. *India and the Knowledge Economy: Leveraging Strengths and Opportunities*. Washington, DC: World Bank.

Dahlman, Carl, Douglas Zhihua Zeng, and S. Wang. 2007. *Building a System of Lifelong Learning to Sharpen China's Competitiveness*. Washington, DC: World Bank.

Dahlman, Carl, and Jean-Eric Aubert. 2001. *China and the Knowledge Economy—Seizing the 21st Century*. WBI Development Series. Washington, DC: World Bank.

Dahlman, Carl, Jorma Routti, Pekka Yla-Anttila, eds. 2006. *Finland as a Knowledge Economy—Elements of Success and Lessons Learned*. World Bank

Dahlman, Carl, and Thomas Andersson, eds. 2000. *Korea and the Knowledge-Based Economy: Making the Transition*. Washington, DC: World Bank.

Dahlman, Carl, and Yevgeny Kuznetsov. 2007. *Mexico's Transition to a Knowledge-Based Competitiveness: Challenges and Opportunities*. Washington, DC: World Bank.

De Ferranti, David, Guillermo E. Perry, Indermit Gill, J. Luis Guasch, William F. Maloney, Carolina Sanchez-Paramo, and Norgert Schady. 2003. "Closing the Gap in Education and Technology." World Bank Latin American and Caribbean Studies, Washington, DC.

Diamond, Jared. 1999. *Guns, Germs and Steel: The Fates of Human Societies*. New York: Norton.

d'Iribarne, Philippe, and Alain Henry. 2003. *Le Tiers monde qui reussit*. Paris: Le Seuil.

Doumbo, Ogobara K. 2005. "Case study of the Malaria Research and Training Center at the University of Bamako, Mali." Paper presented at a Francophone African Universities seminar on lessons from universities' contributions to development, November 30–December 2, Saly, Senegal.

Drucker, Peter. 1996. *The Executive in Action.* New York: HarperCollins.

Dutz, Mark, ed. Forthcoming. *Unleashing India's Innovation Potential.* Washington, DC: World Bank.

Easterly, William, and Ross Levine. 2001. "What Have We Learned from a Decade of Empirical Research on Growth? It's Not Factor Accumulation: Stylized Facts and Growth Models." *World Bank Economic Review* 15: 177.

EFA-FTI Secretariat. 2006: "Status Report Prepared for the Fourth EFA-FTI Partnership Meeting." November 13–14, Cairo.

EIU (Economist Intelligence Unit). 2005. "The Major League: TEDA." April 25.

Elliott, John. 2005. "On the Road to Repair." *Fortune,* October 31.

European Commission. 2003. *Innovation Policy in 7 Candidate Countries: The Challenges.* Brussels.

European Commission. 2004. "European Innovation Scoreboard 2004." http://trendchart.cordis.lu/scoreboards/scoreboard2004.

European Commission. 2005a. "eGovernment in Estonia." Brussels.

European Commission. 2005b. "European Innovation Scoreboard 2005." http://trendchart.cordis.lu/tc_article.cfm?ID=3197&NEWSID=13

Eyzaguirre, N., M. Marcel, J. Rodríguez, and M. Tokman. 2005. "The Way to the Economy of Knowledge: The Road to Growing Equitably in the Long Term." *Estudios Públicos* 97: 5–57, Centro de Estudios Públicos, Santiago, Chile.

Finn, G. 2005. "Stay Rates of Foreign Doctorate Recipients from U.S. Universities, 2003." Oak Ridge Institute for Science and Education, Oak Ridge, Tennessee. November.

Fuentes, Rodrigo, and Verónica Mies. 2005. "Mirando el Desarrollo Económico de Chile: Una Comparación Internacional." Working Papers of the Central Bank of Chile 308. Santiago.

Fujita, Masahisa. 2006. "Economic Development Capitalizing on Brand Agriculture: Turning Development Strategy on Its Head." Paper delivered at the Annual World Bank Conference on Development Economics, Tokyo, May 29–30.

Gaudin, Thierry, and Jean-François Dégremont. 1993. *2100, Récit du prochain siècle.* Paris: Payot.

Girard, Bruce. 2003. "A Passion for Radio." http://www.comunica.org/passion/pdf/chapter23.pdf.

Goldberg, Itzhak, Manuel Trajtenberg, Adam Jaffe, Thomas Muller, Julie Sunderland, and Enrique Blanco Armas. 2006. "Public Financial Support for Commercial Innovation—Europe and Central Asia Knowledge Economy Study." World Bank Report 36064, Washington, DC.

Government of India. 2002. "India Vision 2020." Report of the Planning Commission. New Delhi. http://planningcommission.nic.in/reports/genrep/pl_vsn2020.pdf.

Guellec, Dominique, and Pierre Ralle. 2003. *Les nouvelles theories de la croissance.* Paris: La Decouverte.

Gupta, Anil K. 2006. "From Sink to Source: The Honey Bee Network Documents Indigenous Knowledge and Innovations in India." *Innovations: Technology, Governance, Globalization* 1 (3): 49–66.

Hanushek, Eric A., and Dennis D. Kimko. 2000. "Schooling, Labor-Force Quality, and the Growth of Nations." *American Economic Review* 90(5): 1184–1208. December.

Hausmann, Ricardo, Dani Rodrik, and Andrés Velasco. 2005. *Growth Diagnostics.* John F. Kennedy School of Government, Harvard University, Cambridge, MA.

Heeks, Richard, and Shoba Arun. 2006. "Social Outsourcing as a Development Tool: Outsourcing to Social Enterprises for Poverty Reduction and Women's Empowerment in Kerala." Unpublished paper. http://www.womenictenterprise.org/kasargod.htm.

Holm-Nielsen, Lauritz, and Natalia Agapitova. 2002. "Chile: Science, Technology, and Innovation." LCSHD Paper Series 79, Latin America and the Caribbean Region, World Bank, Washington, DC.

IIDA. 2006. "Ireland, Knowledge Is in Our Nature." Dublin: Irish Industrial Development Agency.

IMF (International Monetary Fund). 2004. "Article IV Consultation: Botswana 2003." Country Report 04/225. Washington, DC.

infoDev. 2006. "ICT, Innovation, and Economic Growth in Transition Economies: A Multi-Country Study of Poland, Russia, and the Baltic Countries." World Bank, Washington, DC. http://www.infodev.org/en/Project.27.html.

Institut d'Economie Quantitative, Ministère du Développement et de la Coopération Internationale, République Tunisienne. 2005. *Rapport Annuel sur l'Economie du Savoir, 2004.* Tunis: Simpact.

Intarakumnerd, Patarapong, Chairatana Pun-arj, and Tipawan Tangchitpiboon. 2002. "National Innovation Systems in Less Successful Developing Countries: The Case of Thailand." *Research Policy* 31 (8–9): 1445–1457.

Kaplan, Robert, and David Norton. 2004. *Strategy Maps: Converting Intangible Assets into Tangible Outcomes.* Cambridge, MA: Harvard Business School Publishing.

Kaplinsky, Raphael. 2005. *Globalization, Inequality, and Poverty: Between a Rock and a Hard Place.* Cambridge, England: Polity Press.

Kikeri, Sunita, Thomas Kenyon, and Vincent Palmade. 2006. *Reforming the Investment Climate—Lessons for Practitioners.* World Bank/IFC, Washington, DC.

Kuznetsov, Yevgeny. 2006. *Diaspora Networks and the International Migration of Skills: How Countries Can Draw on Their Talent Abroad.* WBI Development Studies. Washington, DC: World Bank.

Lall, Sanjaya. 2002. "Science and Technology Policies in Southeast Asia." Proceedings and final reports of the STRATA consolidating workshop on "Science and technology policies in Europe: new challenges, new responses," Brussels, April 22–23 April. http://cordis.europa.eu/improving/strata/workshop.htm.

Land, A. 2002. "Structured Public-Private Sector Dialogue: The Experience of Botswana." ECDPM Discussion Paper 37, EDCPM, Maastricht.

Landes, David S. 1999. *The Wealth and Poverty of Nations: Why Some Are So Rich and Some So Poor.* New York: Norton.

Lanvin, Bruno, and Isabel Neto. 2005. "The Role of International Cooperation in e-Development." In Robert Schware, ed., *E-Development: From Excitement to Effectiveness* (chapter 7). Washington, DC: World Bank.

Lanvin, Bruno, and Christine Zhen-Wei Qiang. 2003. "Poverty e-Readication: Using ICT to Meet MDG—Direct and Indirect Roles of e-Maturity." In Soumitra Dutta, Bruno Lanvin, and Fiona Paua, eds., *The Global Information and Technology Report 2003–2004: Towards an Equitable Information Society* (chapter 3). World Economic Forum. Oxford: Oxford University Press.

Léautier, Frannie. 2006. *Cities in a Globalizing World—Governance, Performance & Sustainability.* Washington, DC: World Bank Institute.

Lederman, Daniel, and Laura Saenz. 2005. "Innovation and Development Around the World: 1960–2000." World Bank Policy Research Working Paper 3774. Washington, DC.

Lederman, Daniel, and William Maloney. 2003. "R&D and Development." World Bank Policy Research Working Paper 3024. Washington, DC.

Leith, J. Clark. 2005. *Why Botswana Prospered.* Montreal: McGill-Queen's University Press.

Levy, Brian, and Sahr Kpundeh, eds. 2004. *Building State Capacity in Africa: New Approaches, Emerging Lessons.* Washington, DC: World Bank.

Lopez-Claros, Augusto, Klaus Schwab, and Michael E. Porter. 2006. *The Global Competitiveness Report 2006–2007: Creating an Improved Business Environment.* New York: Pargrave Macmillan.

Maddison, Angus. 2006. *The World Economy.* Organisation for Economic Co-operation and Development, Paris.

Mamade, Augustinho, Zuber Ahmend, and Peter Easton. 2003. "Mozambique, Repairing the Ravages of War—Initiation Societies and Community Schooling— *IK Notes* 33, Knowledge and Learning Center, Africa Region, World Bank, Washington, DC.

McDermott, Gerald. 2004. "Making a Case for the Argentine Wine Industry." Knowledge@ Wharton. January 22. http://knowledge.wharton.upenn.edu/article.cfm?articleid=904& specialid=13.

McKinsey and Company. 2005. *Building Effective Public-Private Partnerships: Lessons Learnt from the Jordan Education Initiative.* Geneva: World Economic Forum.

McRae, M. 2006. "A Model Diaspora Network: The Origin and Evolution of Global Scot." In Yevgeny Kuznetsov, ed. *Diaspora Networks and the International Migration of Skills: How Countries Can Draw on Their Talent Abroad* (201–220). Washington, DC: World Bank.

Mingat, A., and J. P. Tan. 1996. *Full Social Returns to Education: Estimates Based on Countries Economic Growth Performance.* Washington, DC: World Bank.

Moore, Gordon. 1965. "Cramming More Components into Integrated Circuits." *Electronics Magazine.* April 19.

Mytelka, Lynn K. 2000. "Local Systems of Innovation in a Globalized World Economy." *Industry and Innovation* 7 (1): 15–32.

Ndulu, Benno J., L. Chakraborti, and L. Lijane. 2007. *Challenges of African Growth: Opportunities, Constraints and Strategic Directions.* Africa Region, World Bank, Washington, DC.

Nisbett, Richard. 2003. *The Geography of Thought: How Asians and Westerners Think Differently and Why.* New York: The Free Press.

OECD (Organisation for Economic Co-operation and Development). 1994. *Science, Technology, and Innovation Policies—Federation of Russia.* Paris.

———. 1996. *Science, Technology, and Industry Outlook.* Paris.

———. 1998. *Human Capital Investment: An International Comparison.* OECD Centre for Educational Research and Innovation. Paris.

———. 2002. *Foreign Direct Investment for Development: Maximizing Benefits, Minimizing Costs.* Paris.

———. 2004a. *Internationalisation and Trade in Cross-Border Higher Education.* Paris.

———. 2004b. "Lifelong Learning." OECD Policy Brief. Paris. February.

———. 2004c. "Messages from PISA 2000." Paris. http://www.pisa.oecd.org/dataoecd/ 31/19/34107978.pdf.

———. 2005. *Science, Technology, and Industry Scoreboard 2005.* Paris.

———. 2006a. *Are Students Ready for a Technology-Rich World? What PISA Studies Tell Us.* Paris.

———. 2006b. *Science, Technology and Industry Outlook.* Paris.

———. 2007. *OECD Factbook 2007—Economic, Environmental, and Social Statistics.* Paris http://lysander.sourceoecd.org/vl=3564269/cl=21/nw=1/rpsv/factbook/.

Olson, Mancur. 1982. *The Rise and Decline of Nations: Economic Growth, Stagflation, and Social Rigidities.* New Haven, CT: Yale University Press.

O'Siochru, Sean. 2007. "Effective E-Leadership from the Top: But Will It Be Enough?" Draft report prepared for the World Bank by Nexus Research Cooperative.

Pai, M. A. 2004. "Why IITs Must Be Restructured." September 8. http://in.rediff.com/money/2004/sep/08guest2.htm.

Palmade, Vincent. 2005. "Industry Level Analysis: The Way to Identify the Binding Constraints to Economic Growth." World Bank Policy Research Working Paper 3551. Washington, DC. March.

Rischard, Jean-François. 2002. *High Noon—20 Global Problems and 20 Years to Solve Them.* New York: Basic Books.

Sachs, Jeffery D., and Andrew M. Warner 1995. "Natural Resource Abundance and Economic Growth." NBER Working Papers 5398. National Bureau of Economic Research, Cambridge, MA.

Saint, William. 2006. "Innovation Funds for Higher Education: A Users' Guide for World Bank–Funded Projects." Education Sector Working Paper 1. World Bank, Washington, DC.

Salmi, Jamil. 2004, *Constructing Knowledge Societies.* Washington, DC: World Bank.

Salmi, Jamil, and Arthur M. Hauptman. 2006. "Innovations in Tertiary Education Financing: A Comparative Evaluation of Allocation Mechanisms." Education Sector Working Paper 4. World Bank, Washington, DC.

Saxenian, Annalee. 2000. "Taiwan's Transnational Connections." In M. Aoki, ed., *East Asian Economics and Japanese Industry at a Turning Point.* Tokyo: Research Institute of International Trade and Industry.

———. 2005. "The International Mobility of Entrepreneurs and Regional Upgrading in India and China." UNU-WIDER Project on the International Mobility of Talent, Santiago, Chile. May.

Solow, Robert. 1956. "A Contribution to the Theory of Economic Growth." *Quarterly Journal of Economics* 70 (1): 65–94.

Tokman, Andrea. 2004. "Education and Economic Growth in Chile." Working Papers of the Central Bank of Chile 289, Santiago, Chile.

Transparency International. 2005. Corruption Perceptions Index 2005. http://ww1.transparency.org/cpi/2005/cpi2005.sources.en.html.

UNCTAD (United Nations Conference on Trade and Development). 2005. *World Investment Report.* Geneva.

UNCTAD/ICTSD. 2003. "Intellectual Property Rights, Implications for Development." Policy Discussion Paper, Intellectual Property Rights & Sustainable Development Series. International Centre for Trade and Sustainable Development, Geneva. August 2003. http://www.ictsd.org/iprsonline.

UNDP (United Nations Development Programme). 1999. *Human Development Report 1999.* New York: Oxford University Press.

———. 2005. *Community-Based Networks and Innovative Technologies: New Models to Serve and Empower the Poor.* New York

UNESCO (United Nations Educational, Scientific, and Cultural Organization). 2005. *Towards Knowledge Societies.* Paris.

———. 2006. *EFA Global Monitoring Report 2007.* Paris.

UNIDO (United Nations Industrial Development Office). 2005. *Industrial Development Report 2005: Capacity Building for Catching-up—Historical, Empirical and Policy Dimensions.* Vienna. http://www.unido.org/en/doc/46385.

United Nations Millennium Project. 2005. Investing in Development—A Practical Plan to Achieve the Millennium Development Goals. http://www.unmillenniumproject.org/reports/index_overview.htm.

U.S. Department of State. 2005. "Background Note: Botswana." Bureau of African Affairs, Washington, DC.

Utz, Anuja. 2006. "Fostering Innovation, Productivity, and Technological Change: Tanzania in the Knowledge Economy." World Bank Institute Working Paper. Washington, DC.

Vargas Zuñiga, Fernando. 2005. "Key Competencies and Lifelong Learning." In Fernando Vargas Zuñiga, ed., *Key Competencies and Lifelong Learning* (81–118). Montevideo: CINTERFOR (Inter-American Research and Documentation Center on Vocational Training, International Labour Organization). http://www.ilo.org/public/english/region/ampro/cinterfor/publ/vargas/index.htm.

Vodaphone. 2005. "Africa: The Impact of Mobile Phones." Policy Paper Series 2, Vodaphone, Newbury, Berkshire, England. March. http://www.vodafone.com/start/responsibility/our_social___economic/socio-economic_impact/impact_of_mobile_phones.html.

Watkins, Alfred. 2003. "From Knowledge to Wealth: Transforming Russian Science and Technology for a Modern Knowledge Economy." World Bank Policy Research Working Paper 2974. Washington, DC.

Watson, Robert, Michael Crawford, and Sara Farley. 2003. "Strategic Approaches to Science and Technology in Development." World Bank Policy Research Working Paper 3026. Washington, DC. April.

WBI (World Bank Institute) and KDI (Korea Development Institute). 2007. *Korea as a Knowledge Economy: Evolutionary Process and Lessons Learned.* Washington, DC, and Seoul.

WEF (World Economic Forum). 2003. *Africa Competitiveness Report 2004.* Geneva.

———. 2004. *Global Information Technology Report 2004–2005.* Geneva.

———. 2005. *The Global Competitiveness Report 2005–2006.* Geneva.

———. 2006. *The Global Information Technology Report 2005–2006.* Geneva.

Wellenius, Bjorn. 2002. "Closing the Rural Communications Access Gap: Chile 1995–2002." World Bank Discussion Paper 430. Washington, DC.

Wellenius, Bjorn, C. Rosotto, and A. Lewin. 2004. "Morocco: Developing Competition in Telecommunications." CITPO Working Paper. World Bank, Washington, DC.

Winters, L. Alan, and Shahid Yusuf. 2007. *Dancing with the Giants: China, India, and the Global Economy.* Washington, DC: World Bank.

World Bank. 1993. *The East Asian Miracle.* New York: Oxford University Press.

———. 1999. *World Development Report 1998/99: Knowledge for Development.* Oxford University Press.

———. 2000a. "Botswana: An Example of Prudent Economic Policy and Growth." Africa Region Findings 161. Washington, DC. June. http://www.worldbank.org/afr/findings/english/findtoc.htm.

———. 2000b. *Can Africa Claim the 21st Century?* Washington, DC.

———. 2000c. *Higher Education in Developing Countries: Peril and Promise.* Washington, DC.

———. 2000d. "Colombia: Rural Education Project." Project Appraisal Document, Report 20099, Latin America and the Caribbean Region. Washington, DC.

———. 2001. *Global Development Finance 2001.* Washington, DC.

———. 2002a. "Constructing Knowledge Societies: New Challenges for Tertiary Education." Education Group, Human Development Network. Washington, DC.

———. 2002b. *From Natural Resources to the Knowledge Economy: Trade and Job Quality.* Latin America and Caribbean Region. Washington, DC.

———. 2003a. *Closing the Gap in Education and Technology.* Latin America and the Caribbean Region. Washington, DC.

———. 2003b. "ICT and MDGs: A World Bank Group Perspective." Working Paper 27877. Washington, DC.

———. 2003c. *Information and Communication Technologies, Poverty and Development: Learning from Experience.* Washington, DC.

———. 2003d. *Lifelong Learning in the Global Knowledge Economy: Challenges for Developing Countries.* New York: Oxford University Press.

———. 2003e. "Mauritania: Enhanced National Capacity in Telecommunications Sector Reforms." Africa Region Findings 225. Washington, DC. April. http://www.worldbank.org/afr/findings/english/findtoc.htm.

———. 2003f. "Mozambique: Improving the Quality of University Graduates." Africa Region Findings. Washington, DC. July.

———. 2003g. "Technology, Skills and Internet Services in Korea: Moving Towards a Knowledge-based Economy." Report 23905-KO. Washington, DC. June.

———. 2004a. "Building Blocks of e-Government: Lessons from Developing Countries." PREM Note 91. August. Washington, DC.

———. 2004b. *Doing Business in 2005: Removing Obstacles to Growth.* Washington, DC.

———. 2004c. "India: Investment Climate Assessment 2004: Improving Manufacturing Competitiveness." Finance and Private Sector Development Unit, South Asia Region, Washington, DC.

———. 2004d. *Indigenous Knowledge: Local Pathways to Global Development.* Africa Region, Washington, DC.

———. 2004e. "Innovation Systems: World Bank Support of Science and Technology Development." World Bank Working Paper 32. Washington, DC.

———. 2004f. *Scaling Up Poverty Reduction.* Report of "Reducing Poverty, Sustaining Growth: A Global Learning Process and Conference," Shanghai, May 25–27. Washington, DC: World Bank. http://info.worldbank.org/etools/reducingpoverty/index.html.

———. 2004g. "Republic of Tunisia: Review of Development Policies." Economic and Social Development Group, Middle East and North Africa Region. Washington, DC. October.

———. 2004h. *World Development Report 2004: Making Services Work For Poor People.* Washington, DC.

———. 2005a. *Doing Business in 2006: Creating Jobs.* Washington, DC.

———. 2005b. *E-Development: From Excitement to Effectiveness.* Washington, DC.

———. 2005c. "Education Sector Strategy Update: Achieving Education for All, Broadening Our Perspective, Maximizing Our Effectiveness." Report Number 34850. Washington, DC.

———. 2005d. *E-Strategies Monitoring and Evaluation Toolkit.* Washington, DC.

———. 2005e. *Expanding Opportunities and Building Competencies for Young People: A New Agenda for Secondary Education.* Washington, DC.

———. 2005f. "Vietnam: Project Appraisal Document for an Information and Communication Technology Development Project." Report 32919, Global Information and Communication Technologies Department, East Asia and Pacific Region. Washington, DC. August 12.

———. 2005g. "Public-Private Sector Dialogue in Uganda's Reform Process." Africa Region Findings 253. August. http://www.worldbank.org/afr/findings/english/findtoc.htm.

———. 2005h. *World Development Report 2005: A Better Investment Climate for Everyone*. Washington, DC.

———. 2006a. "Bangladesh: End of MFA Quotas—Key Issues and Strategic Options for Bangladesh Readymade Garment Industry." Report 34964-BD. Washington, DC. January 18.

———. 2006b. "Country Assistance Strategy for Bangladesh, FY06–09." Report 35193. Washington, DC. February 2.

———. 2006c. *Doing Business 2007: How to Reform*. Washington, DC: World Bank.

———. 2006d. "From Schooling Access to Learning Outcomes: An Unfinished Agenda, An Evaluation of World Bank Support to Primary Education." Independent Evaluation Group. Washington, DC.

———. 2006e. *Global Integration and Technology Transfer*. Washington, DC.

———. 2006f. *Information and Communication for Development: Global Trends and Policies*. Washington, DC.

———. 2006g. "Lessons from Best Practices in Promoting Education for All." Prepared by the World Bank Institute, Human Development Unit, for a conference on Latin America and the Caribbean, Cartagena de Indias, October.

———. 2006h. "Public Financial Support for Commercial Innovation." ECA Knowledge Economy Study, Part I, Private and Financial Sector Development Department. Europe and Central Asia Region. Washington, DC.

———. 2007. "Innovation Policies in Latin America II." Office of the Chief Economist, Latin America and the Caribbean Region. Washington, DC.

Yusuf, Shahid. 2003. *Innovative East Asia: The Future of Growth*. New York: Oxford University Press.

Yusuf, Shahid, M. Anjum Altaf and Kaoru Nabeshima, eds. 2004a. Global Change and East Asian Policy Initiatives. New York: Oxford University Press.

———. 2004b. *Global Production Networking and Technological Change in East Asia*. New York: Oxford University Press.

Yusuf, Shahid, and Kaoru Nabeshima. 2006. *Postindustrial East Asian Cities: Innovation for Growth*. Washington, DC: World Bank.

Yusuf, Shahid, and Simon J. Evenett. 2002. *Can East Asia Compete?* New York: Oxford University Press.

Zagha, Roberto, and Gobind Nankani, ed. 2006. *Economic Growth in the 1990s: Learning from a Decade of Reform*. Washington, DC: World Bank.

Zeng, Douglas Zhihua. 2007. *Knowledge, Technology and Cluster-Based Growth in Africa: Finding from Eleven Case Studies of Enterprise Clusters in Africa*. Washington, DC: World Bank Institute.

Zhen-Wei Qiang, Christine, Alexander Pitt, and Seth Ayers. 2004. *Contribution of Information and Communication Technologies to Growth*. Washington, DC: World Bank.